Table of Contents

Introduction	1
Metals: Drude model	
Charge transport/DC conductivity	4
Hall effect	6
Noise	8
AC conductivity	9
Thermal properties	14
Metals: Sommerfeld model	16
Boltzmann charge transport	19
Thermionic emission	21
Coherent transport/current-voltage	22
Quantum of conductance	26
Coulomb blockade	31
Quantum Hall effect	32
Thermodynamics	35
Phonons	39
Interactions	
Electrostatic screening	45
Ferromagnetism	47
Spin injection	55
Bandstructure	
Nearly-free electron/Fourier basis	59
Bare essentials of group theory	71
LCAO/tight binding	77
Semiconductors	88
Homework assignments and solutions	95

What is Solid-State Physics?

Our goal is to understand how to model macroscopic physical properties of solids conferred by thermodynamically-large ensembles of their constituent microscopic systems

"Models":
- isotropic "gas" (continuous symmetry)
- ordered lattices (periodic/ discrete symmetry)
- amorphous ("no" symmetry)

- interacting: (additional terms in Hamiltonian)
- noninteracting: (statistics)

"Physical Properties":
- *linear response*: Change X, measure proportional change in Y, where "property" is proportionality constant.

- *spectroscopic*: Absolute value of Y not as important as (nonlinear) functional form & critical values of X.

Linear Response Examples

	"X"	"Y"	Proportionality const	typical value @ RT
"Ohm's"	gradient of electrostatic potential $\vec{\nabla}\phi$	charge current density $\vec{J}$	electrical conductivity σ	$10^{-18} < \sigma < 10^{6}$ [S/cm] (quartz – Ag)
"Fourier's"	gradient of temperature $\vec{\nabla}T$	heat flux $\vec{J}_Q$	thermal conductivity κ	$1 < \kappa < 10^{3}$ [W/Kcm] (glass – diamond)
	change in energy ΔU	change in temperature ΔT	heat capacity C_V	~ 1 [J/Kcm³]
	magnetic field $\vec{H}$	magnetization $\vec{M}$	magnetic susceptibility χ	$1/\chi$ =0 "ferro", >0 "para", <0 "dia"

Spectroscopic Examples

Optical

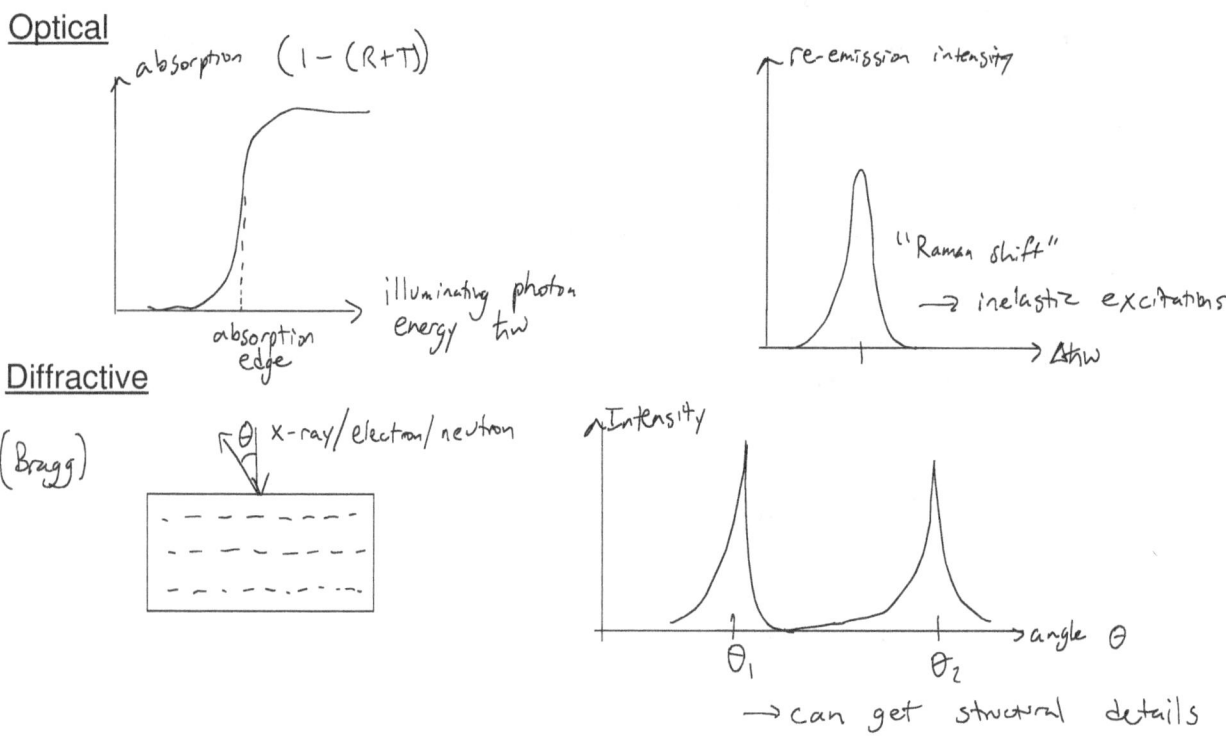

Diffractive

(Bragg)

Electron Transport ("device physics")

"local"

"nonlocal"

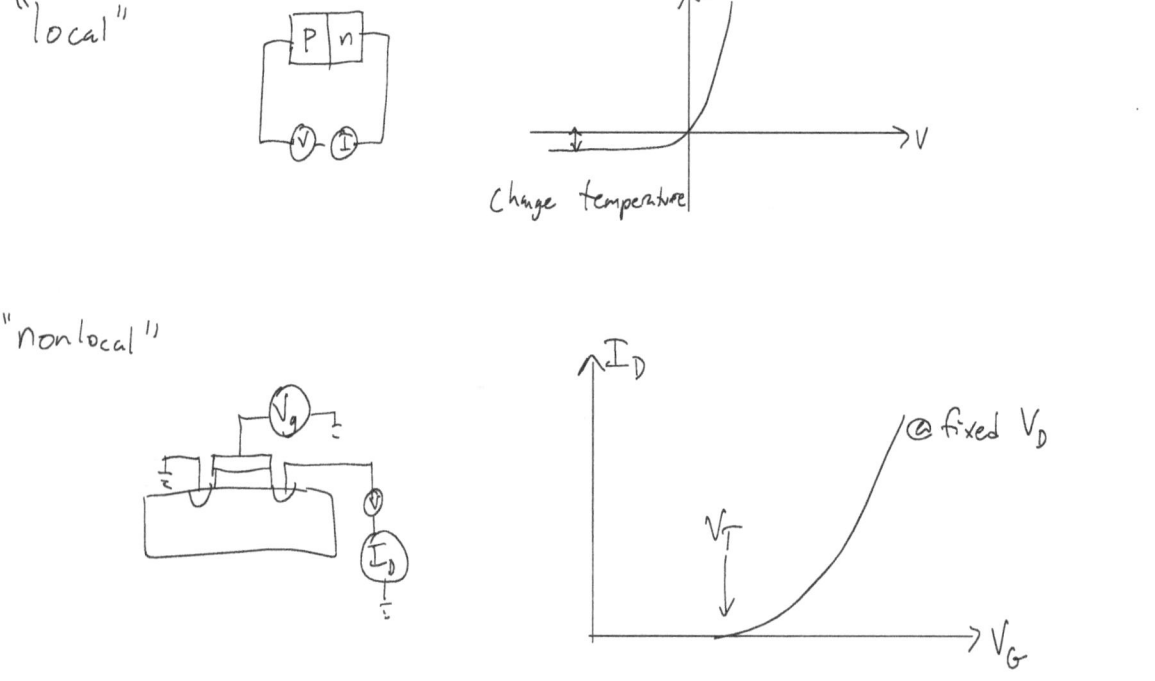

Magnetic Field

ferromagnetic hysteresis

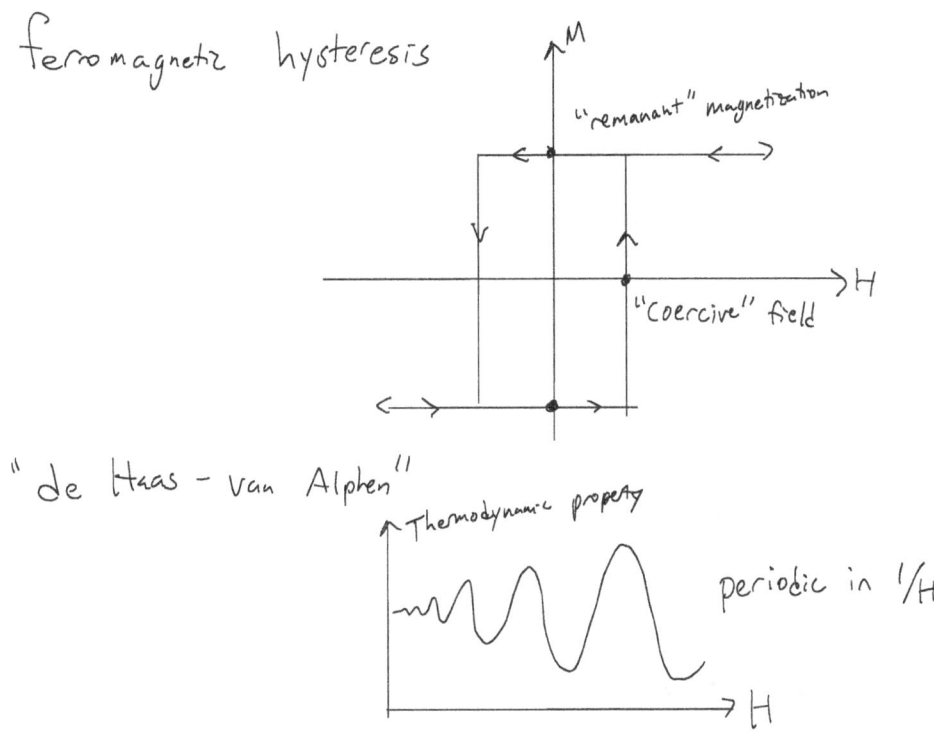

"de Haas – van Alphen"

periodic in $1/H$

Temperature

- heat capacity: ~const. at high temps, $\propto T^3$ @ low temps
  $\propto T$ @ <u>very</u> low temps

- electrical conductivity

  $1/\sigma = \rho$ "resistivity"
  "insulator"
  "metal"
  RT

- Susceptibility $|1/\chi|$
  para/dia   ferro
  T_C "critical" temperature – phase transition

Outline of Course Topics

Metals: Drude model
    Charge transport/DC conductivity
    Hall effect
    Noise
    AC conductivity
    Thermal properties
Metals: Sommerfeld model
    Boltzmann charge transport
    Thermionic emission
    Coherent transport/current-voltage
    Quantum Hall effect
    Thermodynamics
Phonons
Interactions
    Electrostatic screening
    Ferromagnetism
    Spin injection
Bandstructure
    Nearly-free electron/Fourier basis
    Bare essentials of group theory
    LCAO/tight binding
Semiconductors

Classical Model of Metals

"In the beginning, there was...."

Drude's ideal electron gas (1900):

In equilibrium $|\vec{E}| = 0$ $\vec{E} \neq 0$ ←

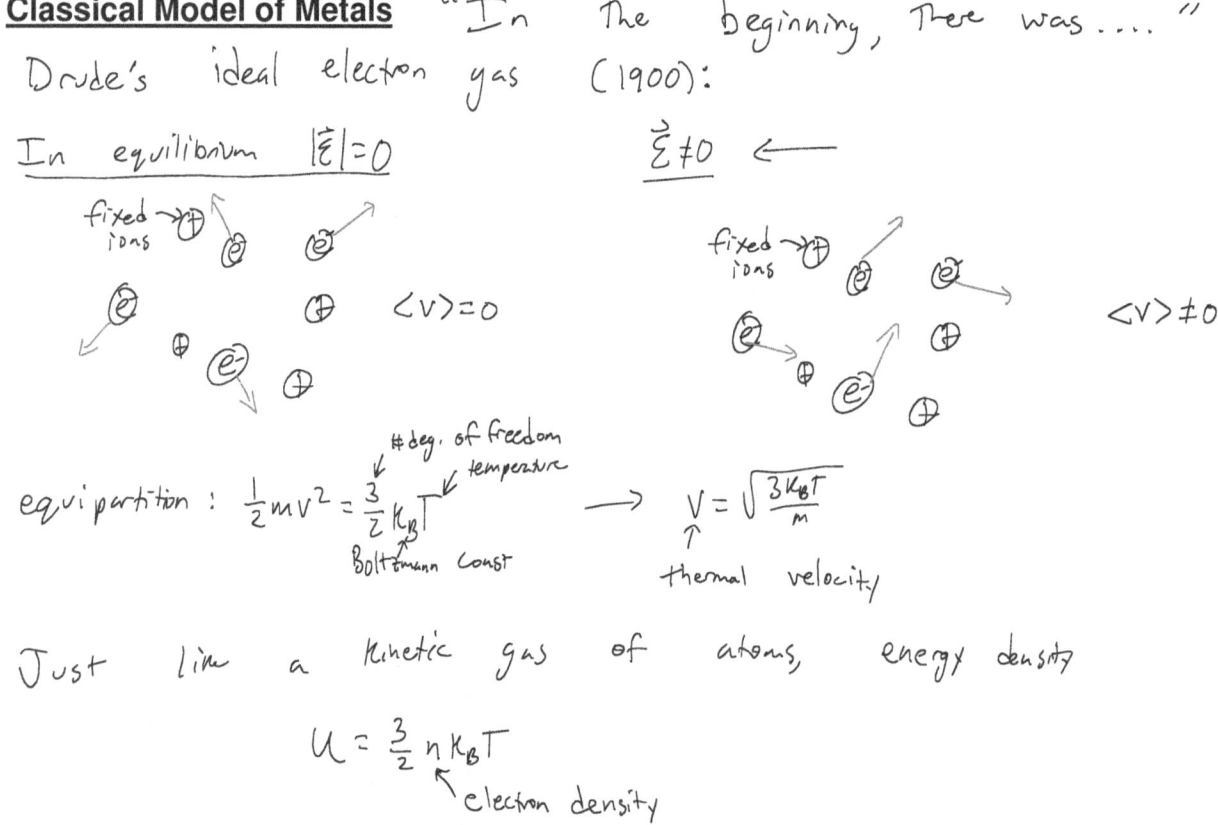

equipartition: $\frac{1}{2}mv^2 = \frac{3}{2} k_B T$ → $v = \sqrt{\frac{3 k_B T}{m}}$

(# deg. of freedom, temperature, Boltzmann const)   ↑ thermal velocity

Just like a kinetic gas of atoms, energy density

$$U = \frac{3}{2} n k_B T$$

↖ electron density

Equation of Motion ($\vec{\mathcal{E}} \neq 0$)

Newton's 2nd Law

$$\dot{p} = F \longrightarrow \frac{dp}{dt} = \underbrace{-e\mathcal{E}}_{\text{Coulomb}} - \underbrace{\frac{p}{\tau}}_{\text{Collision}} \text{ "scattering time"}$$

momentum relaxation

For const. $\mathcal{E}$ (DC) in steady-state

$$p = -e\mathcal{E}\tau \quad \text{so} \quad v_D = \frac{p}{m} = -\frac{e\tau}{m}\mathcal{E} = \mu\mathcal{E} \quad \text{"mobility"} \quad \left[\frac{cm/s}{V/cm} = \frac{cm^2}{Vs}\right]$$

Charge current density (charge flux)

$$j = -env_D = -ne\mu\mathcal{E} = \sigma_0 \mathcal{E}$$

$$\sigma_0 = \frac{ne^2\tau}{m} \quad \text{"DC Drude Conductivity"}$$

Interpretation / Quantitative approximation

- Density n: ~1 from each atom so $n \sim 10^{23}/cm^3$ (from Avogadro's number)

- Scattering time τ: for a "good" conductor @ RT $\sigma \sim 10^6 \, S/cm$

$$\sigma = \frac{ne^2\tau}{m} \rightarrow \tau = \frac{m\sigma}{ne^2}$$

$[\sigma]: \frac{S}{cm} = \frac{1}{\Omega cm} = \frac{C/s}{V \cdot cm} = \frac{e}{1.6 \times 10^{-19} \, V \cdot cm \cdot s} = \frac{e^2}{1.6 \times 10^{-19} \, eV \cdot cm \cdot s}$

$[m]: 5.11 \times 10^5 \, eV/c^2$ (electron mass)

$$\tau = \frac{5 \times 10^5 \, eV \cdot s^2}{9 \times 10^{20} \, cm^2} \cdot 10^6 \cdot \frac{e^2}{1.6 \times 10^{-19} \, eV \cdot cm \cdot s} \sim 10^{-14} \, s \quad (10 \, fs!)$$
$$\overline{10^{23} \, cm^{-3} \cdot e^2}$$

Very short timescale, cannot be measured experimentally. Can convert to accessible lengthscale though...

Mean-free path (Scattering length)

We expect scattering from positive background ions:

$\lambda = v\tau$

$v = \sqrt{\frac{3 k_B T}{m}}$ @ RT (300K) $k_B T \sim \frac{1}{40}$ eV

$= \sqrt{\frac{3 \cdot \frac{1}{40} eV \cdot 9 \times 10^{20} cm^2}{5 \times 10^5 eV \cdot s^2}} \sim 10^7 \frac{cm}{s}$

So $\lambda \cong 10^7 \frac{cm}{s} \cdot 10^{-14} s \sim 10^{-7} cm = 1 nm$

This is on the order of atomic length scales so seems reasonable. However, is it reasonable to explain lower conductivities with much shorter λ? Is τ just a fudge factor? We need a prediction of theory that is independent of τ!

Hall effect

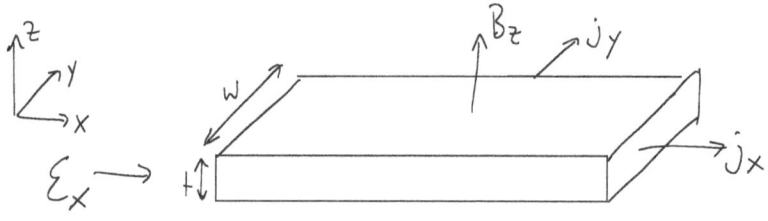

Eqn. of motion: $\frac{d\vec{p}}{dt} = -e\left(\vec{\mathcal{E}} + \frac{\vec{p}}{m} \times \vec{B}\right) - \frac{\vec{p}}{\tau}$

Vector components:

$\hat{x}: \dot{p}_x = -e\mathcal{E}_x - \frac{eB_z}{m} p_y - \frac{p_x}{\tau}$

$\hat{y}: \dot{p}_y = -e\mathcal{E}_y + \frac{eB_z}{m} p_x - \frac{p_y}{\tau}$

Units $\left[\frac{eB}{m}\right] = \frac{1}{time}$

define $\omega_c = \frac{eB_z}{m}$

In steady-state

$\sigma_0 \mathcal{E}_x = \omega_c \tau j_y + j_x$

$\sigma_0 \mathcal{E}_y = -\omega_c \tau j_x + j_y$

Resistivity and conductivity tensors

$$\begin{bmatrix} \mathcal{E}_x \\ \mathcal{E}_y \end{bmatrix} = \frac{1}{\sigma_0} \begin{bmatrix} 1 & \omega_c \tau \\ -\omega_c \tau & 1 \end{bmatrix} \begin{bmatrix} j_x \\ j_y \end{bmatrix} \rightarrow \vec{\mathcal{E}} = \overleftrightarrow{\rho} \vec{j}, \text{ where } \overleftrightarrow{\rho} = \begin{bmatrix} \rho_{xx} & \rho_{xy} \\ -\rho_{xy} & \rho_{yy} \end{bmatrix}$$

$$\vec{j} = \overleftrightarrow{\sigma} \vec{\mathcal{E}}, \text{ where } \overleftrightarrow{\sigma} = \overleftrightarrow{\rho}^{-1} = \frac{\sigma_0}{1+\omega_c^2 \tau^2} \begin{bmatrix} 1 & -\omega_c \tau \\ \omega_c \tau & 1 \end{bmatrix} = \begin{bmatrix} \sigma_{xx} & -\sigma_{xy} \\ \sigma_{xy} & \sigma_{yy} \end{bmatrix}$$

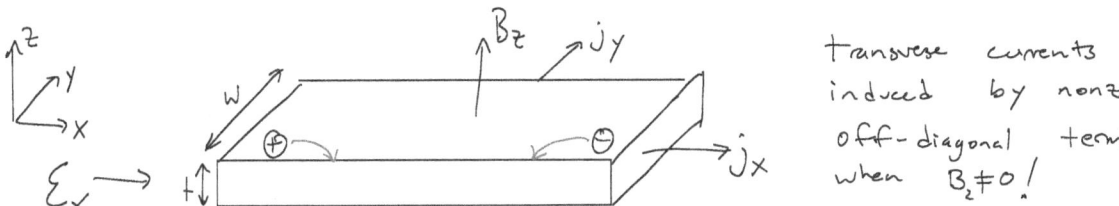

transverse currents induced by nonzero off-diagonal terms when $B_z \neq 0$!

If the transverse edges are unconnected, charge will accumulate until Coulomb and Lorentz forces balance in equilibrium!

Open circuit ($j_y = 0$, charge accumulation)

$$\sigma_0 \mathcal{E}_y = -\omega_c \tau j_x + \cancel{j_y}^{\,0}$$

$$\mathcal{E}_y = \frac{m}{ne^2 \cancel{\tau}} \cdot \left(-\frac{eB}{m}\right)\cancel{\tau} j_x = -\frac{B}{ne} j_x = \rho_{xy} j_x$$

Empirically, $V_{Hall} = V_y = \mathcal{E}_y W = -\frac{B}{ne} \frac{I_x}{t \cancel{W}} \cancel{W} = -\frac{B_z}{net} I_x$ "Hall resistance" (no τ!)

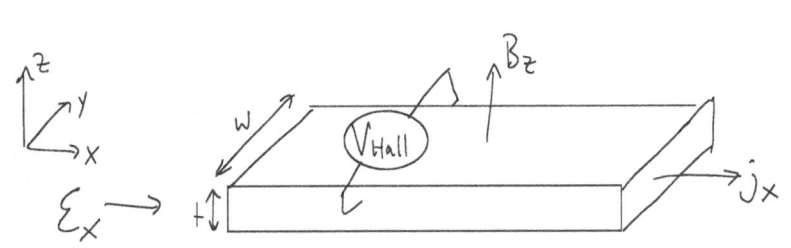

- get sign and density n from slope of $R_{Hall} = -\frac{B_z}{net}$
- get mobility (and τ) from conductivity $\sigma_{xx} = ne\mu$

Fluctuation from random motion

$v(t)$ time series of an individual electron's motion

autocorrelation fn: $C_{vel}^i(t) = \langle \delta v_i(t_0) \cdot \delta v_i(t_0+t) \rangle$

$= \langle \delta v_i^2 \rangle e^{-\frac{|t|}{\tau}}$

equipartition $\frac{1}{2} m v^2 = \frac{1}{2} k_B T$ gives $\langle \delta v_i^2 \rangle = \frac{k_B T}{m} \rightarrow C_{vel}^i(t) = \frac{k_B T}{m} e^{-\frac{|t|}{\tau}}$

power-spectral density can be recovered via "Wiener-Khinchine" Thm:

$$S_{vel}^i = 2\int_{-\infty}^{\infty} C_{vel}^i(t) e^{-i\omega t} dt = \frac{2 k_B T}{m}\left[\int_{-\infty}^{0} e^{-(i\omega - \frac{1}{\tau})t} dt + \int_0^{\infty} e^{-(i\omega + \frac{1}{\tau})t} dt\right]$$

$$= \frac{2 k_B T}{m}\left[\frac{1}{-(i\omega - \frac{1}{\tau})} + \frac{-1}{-(i\omega + \frac{1}{\tau})}\right]$$

$$= \frac{2 k_B T}{m} \frac{2\tau}{1 + \omega^2 \tau^2} \qquad (\text{velocity PSD})$$

Voltage power spectral density

$[S_{vel}] = \left[\frac{k_B T \tau}{m}\right] = \frac{eV}{\frac{eV \cdot s^2}{cm^2}} s = \frac{Vel^2}{Hz}$ (power per frequency interval)

Since $V = IR = \frac{ev}{L} \cdot R = \frac{eR}{L} \cdot v$

Current due to one electron

$S_{voltage} = \left(\frac{eR}{L}\right)^2 S_{vel}^i \cdot N^{\text{\# of electrons}} = 4 k_B T \frac{\tau}{m} \cdot \frac{1}{1+\omega^2\tau^2} \cdot \frac{R^2 e^2}{L^2} \cdot A \cdot L \cdot n^{\text{density}}$

(→ 1 for $\omega \ll \frac{1}{\tau}$)

$= 4 k_B T \left(\frac{n e^2 \tau}{m} \frac{A}{L}\right) R^2 = 4 k_B T R \qquad \text{units of } \frac{Volts^2}{Hz}$

"Noise" $= \sqrt{PSD} = \sqrt{4 k_B T R} \quad \left(\frac{Volts}{\sqrt{Hz}}\right) \qquad$ Nyquist-Johnson noise

Voltage Noise

Example: for $R = 50\Omega$, $T = 300K$

Voltage noise: $\sqrt{4 k_B T R} \sim \sqrt{4 \cdot \frac{1}{40} eV \cdot 50\Omega} = \sqrt{10^{-1} eV \cdot 50 \cdot \frac{1.6 \times 10^{-19} V}{e \cdot Hz}}$

Note: $\Omega = \frac{V}{A} = \frac{V \cdot s}{C} = \frac{1.6 \times 10^{-19} V}{e \cdot 1 Hz}$

$\approx 10^{-9} \frac{V}{\sqrt{Hz}} = 1 nV/\sqrt{Hz}$

This noise will limit our ability to measure small voltages in the lab @ RT!

Solution in time-varying E-field

$\mathcal{E} \to Re\{\mathcal{E}(\omega) \exp(-i\omega t)\} \xrightarrow{\text{linear response}} p = Re\{p(\omega) \exp(-i\omega t)\}$

Eqn of motion: $\frac{dp}{dt} = -e\mathcal{E} - \frac{p}{\tau} \to -i\omega p(\omega) = -e\mathcal{E}(\omega) - \frac{p(\omega)}{\tau}$

$p(-i\omega + \frac{1}{\tau}) = -e\mathcal{E} \to p = \frac{-e\tau}{1-i\omega\tau}\mathcal{E}$ So $v = \frac{(-\frac{e\tau}{m})}{1-i\omega\tau}\mathcal{E}$

$\underbrace{\phantom{\frac{(-\frac{e\tau}{m})}{1-i\omega\tau}}}_{\text{AC mobility}}$

$\sigma = ne\mu_{AC} = \frac{ne\mu_{DC}}{1-i\omega\tau} = \frac{\sigma_0}{1-i\omega\tau}$

Regimes: $\omega \ll \frac{1}{\tau}$ (up to ~THz) $\sigma(\omega) \to \sigma_0$ (dissipative)

$\omega \gg \frac{1}{\tau}$ (optical) $\sigma(\omega) \to \frac{i\sigma_0}{\omega\tau}$ ("collisionless plasma")

An imaginary conductivity? What does that mean?

Kinetic Inductance

Previously, in DC response, we construct resistance $R = \frac{L}{\sigma_0 A}$.

Now, in AC response, we have an impedance

$$Z = \frac{L}{\sigma(\omega) A} = \frac{L}{\sigma_0 A}(1 - i\omega\tau)$$

The term proportional to $i\omega$ acts as an inductance and is caused by the electron's inertia!

This "Kinetic inductance" is typically very small @ low frequency, but is especially important when R vanishes in a superconductor.

Wave propagation in metals

Faraday's Law: $\vec{\nabla} \times \vec{E} = -\frac{\partial \vec{B}}{\partial t}$

Ampere's Law: $\vec{\nabla} \times \vec{B} = \mu\left(\vec{J} + \epsilon \frac{\partial \vec{E}}{\partial t}\right)$

(permeability, permittivity)

Curl of Faraday's Law: $\vec{\nabla} \times \vec{\nabla} \times \vec{E} = \vec{\nabla}(\vec{\nabla} \cdot \vec{E}) - \nabla^2 \vec{E} = -\frac{\partial}{\partial t}(\vec{\nabla} \times \vec{B})$

(0 by Gauss' Law)

$$-\nabla^2 \vec{E} = -\frac{\partial}{\partial t}\left(\mu\sigma \vec{E} + \mu\epsilon \frac{\partial \vec{E}}{\partial t}\right)$$

Plane wave solutions $\vec{E} = \vec{E}(\omega) \exp(-i\omega t)$

$$-\nabla^2 \vec{E} = \mu\sigma i\omega \vec{E} + \omega^2 \mu\epsilon \vec{E} = \omega^2 \mu\epsilon\left(1 + \frac{i\sigma}{\omega\epsilon}\right)\vec{E}$$

Same as wave eqn in dielectric, except $\epsilon \rightarrow \epsilon\left(1 + \frac{i\sigma}{\omega\epsilon}\right)$

High-frequency regime $(\omega \gg \frac{1}{\tau})$

$$\sigma \to \frac{i\sigma_0}{\omega\tau} \quad \text{so} \quad \epsilon \to \epsilon\left(1 - \frac{\sigma_0}{\epsilon\tau\omega^2}\right) = \epsilon\left(1 - \frac{\omega_p^2}{\omega^2}\right)$$

where "plasma frequency" $\omega_p = \sqrt{\frac{\sigma_0}{\epsilon\tau}} = \sqrt{\frac{ne^2\tau/m}{\epsilon\tau}} = \sqrt{\frac{ne^2}{\epsilon m}}$

Since index of refraction $\tilde{n} = \sqrt{\epsilon(\omega)}$, E field e^{ikx} with dispersion $\omega = \frac{kc}{\sqrt{\epsilon(\omega)}}$ is exponentially suppressed for $\epsilon < 0$

[Graph: $\epsilon(\omega)$ vs ω, showing $\epsilon<0$ damped region and $\epsilon>0$ propagating region, asymptote at ϵ, crossing zero at ω_p]

How big is ω_p?

Estimating plasma frequency

ideal metal, $n \sim 10^{23}/cm^3$

$$\omega_p = \sqrt{\frac{ne^2}{m\epsilon}} = \sqrt{\frac{10^{23}/cm^3}{5\times 10^5 eV/c^2} \frac{e^2}{\epsilon}}$$

we need $\frac{e^2}{\epsilon}$ in units of eV cm!

Since $\frac{e^2}{4\pi\epsilon_0 \hbar c} = \alpha \sim \frac{1}{137}$ "fine structure constant"

$$\frac{e^2}{\epsilon_0} = 2hc\alpha \sim \frac{2 \cdot 1240 \, nm\, eV}{137} \sim 18 \, nm\, eV = 1.8 \times 10^{-6} \, eV\, cm$$

So $\omega_p = \sqrt{\frac{10^{23}/cm^3 \cdot 10^{-6} eV\, cm \cdot 10^{21} cm^2/s^2}{10^5 eV \quad \epsilon_{rel}}} \quad \xrightarrow{\epsilon_{rel} \to 1} \quad 10^{16} \, s^{-1} \gg \frac{1}{\tau}$

$\hbar\omega \sim 10 eV$ (ultraviolet)
"ultraviolet transparency"

What happens at interface between vacuum and metal?

Surface reflection

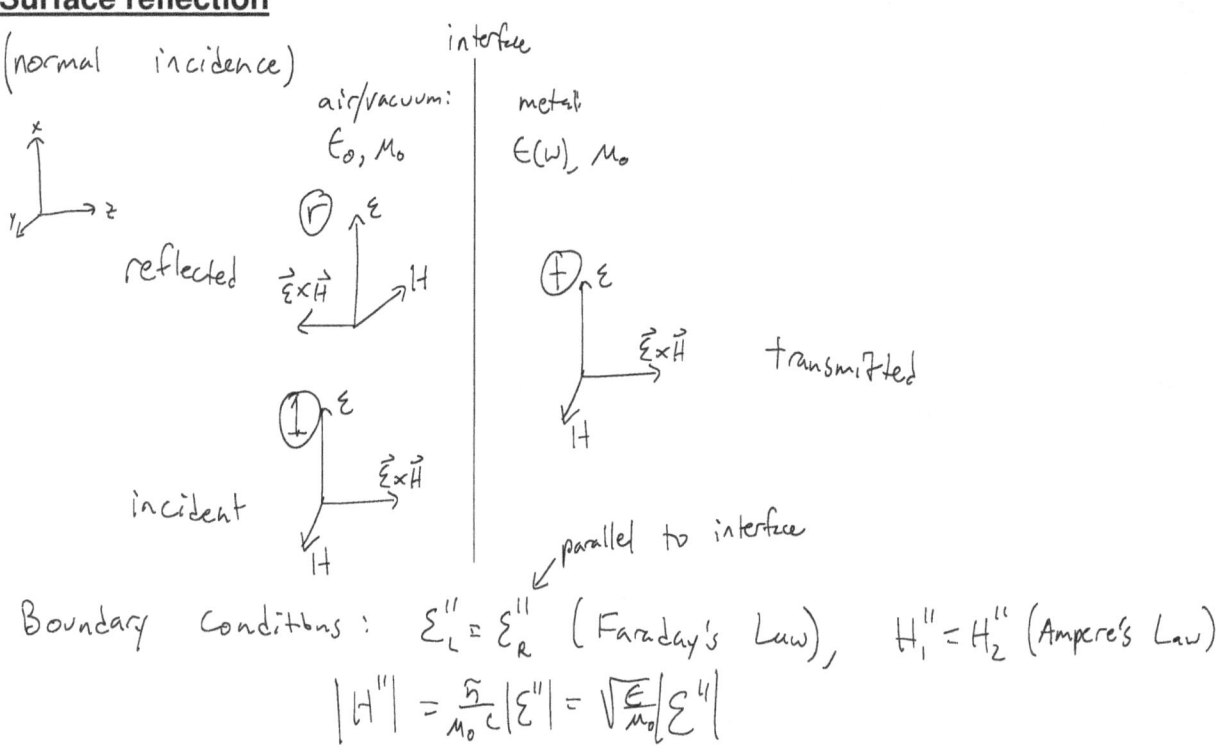

(normal incidence)

air/vacuum: ε_0, μ_0 | metal: $\varepsilon(\omega), \mu_0$

Boundary conditions: $\mathcal{E}_L^{\parallel} = \mathcal{E}_R^{\parallel}$ (Faraday's Law), $H_1^{\parallel} = H_2^{\parallel}$ (Ampere's Law)

$$|H^{\parallel}| = \frac{n}{\mu_0 c}|\mathcal{E}^{\parallel}| = \sqrt{\frac{\varepsilon}{\mu_0}}|\mathcal{E}^{\parallel}|$$

Reflectance

$1 + r = t$ (Faraday's Law) $(1-r)\sqrt{\varepsilon_0} = t\sqrt{\varepsilon(\omega)}$ (Ampere's Law)

$$(1-r)\sqrt{\varepsilon_0} = (1+r)\sqrt{\varepsilon(\omega)} \longrightarrow r = \frac{\sqrt{\varepsilon_0} - \sqrt{\varepsilon(\omega)}}{\sqrt{\varepsilon_0} + \sqrt{\varepsilon(\omega)}} \xrightarrow{\omega \gg \frac{1}{\tau}} \frac{1 - \frac{\sqrt{\varepsilon}}{\sqrt{\varepsilon_0}}\sqrt{1 - \frac{\omega_p^2}{\omega^2}}}{1 + \frac{\sqrt{\varepsilon}}{\sqrt{\varepsilon_0}}\sqrt{1 - \frac{\omega_p^2}{\omega^2}}}$$

$R = |r|^2$ "Reflectance" (ratio of reflected to incident intensity)

Example: $\varepsilon = \varepsilon_0$ $\sqrt{1 - \frac{\omega_p^2}{\omega^2}} = A + iB$ ($\omega > \omega_p$ / $\omega < \omega_p$)

$$R = \left|\frac{1-(A+iB)}{1+(A+iB)}\right|^2 = \frac{(1-A)^2 + B^2}{(1+A)^2 + B^2} \longrightarrow \begin{cases} 1, A=0 \; (\omega < \omega_p) \\ \to 0, B=0 \; (\omega > \omega_p, \omega \to \infty) \end{cases}$$

perfect reflection for $\omega < \omega_p$
So, metals are "lustrous" (shiny)

General response $\left(\sigma = \sigma_0/(1-i\omega\tau),\ \epsilon = \epsilon_{rel}\cdot\epsilon_0\right)$

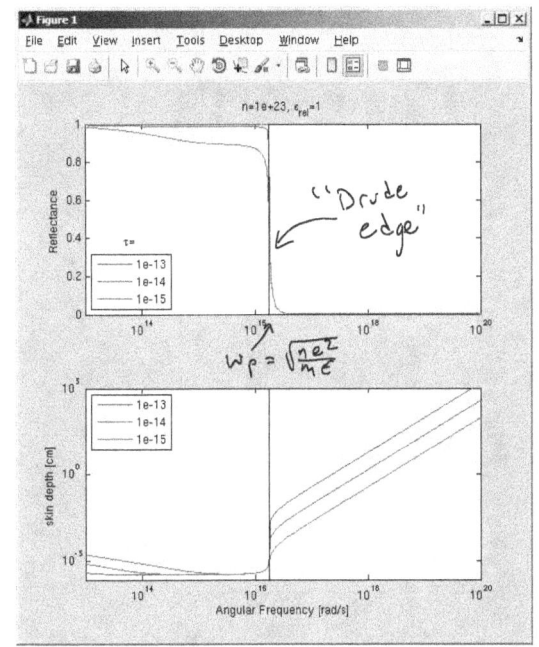
"Drude edge" ← $\omega_p = \sqrt{\frac{ne^2}{m\epsilon}}$

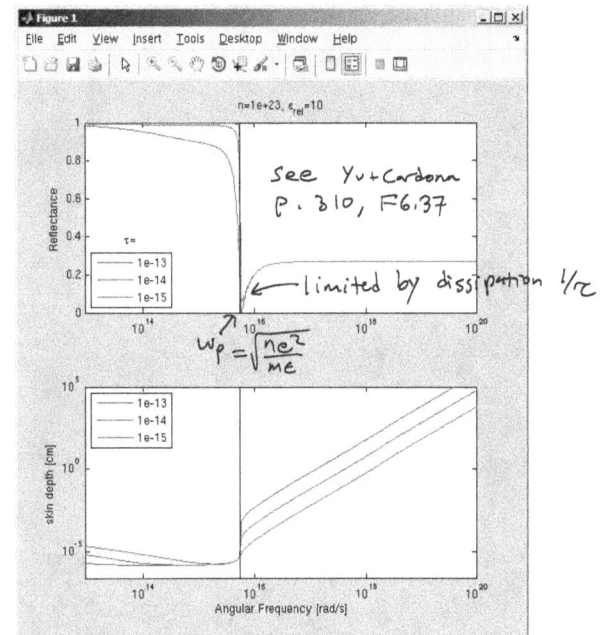
see Yu + Cardona p. 310, F 6.37
← limited by dissipation $1/\tau$
$\omega_p = \sqrt{\frac{ne^2}{m\epsilon}}$

Charge density oscillation (plasmon)

Continuity equation: $\vec{\nabla}\cdot\vec{j} = -\frac{\partial \rho}{\partial t}$ ← charge density
  $\vec{j}$ = current density
  $\rho(t) = \text{Re}\{\rho(\omega)e^{-i\omega t}\}$ ⟹ $\vec{\nabla}\cdot\vec{j} = i\omega\rho(\omega)$

By Ohm's Law, $\vec{\nabla}\cdot\vec{j} = \vec{\nabla}\cdot\sigma\vec{E} = \sigma(\vec{\nabla}\cdot\vec{E}) = \frac{\sigma\rho(\omega)}{\epsilon}$

Substitution: $i\omega\rho = \frac{\sigma}{\epsilon}\rho \rightarrow \omega = \frac{-i\sigma}{\epsilon} \xrightarrow{\omega\gg 1/\tau} \frac{-i(i\sigma_0/\omega\tau)}{\epsilon} = \frac{ne^2\tau/m}{\epsilon\omega\tau}$

$\omega^2 = \frac{ne^2}{m\epsilon} = \omega_p^2$

Newton's 2nd Law:

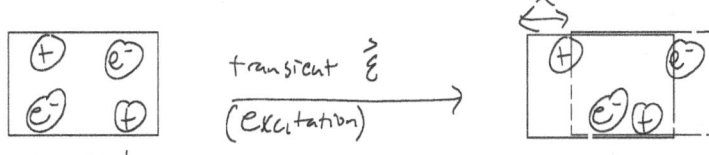
transient $\vec{E}$ (excitation)

response: ← from dipole
$m\ddot{x} = -\vec{E}e = -2\frac{nxe}{2\epsilon}e = -\frac{ne^2}{\epsilon}x \longrightarrow x(t) \propto e^{i\sqrt{\frac{ne^2}{m\epsilon}}t} = e^{i\omega_p t}$

Thermal conductivity: 1D heat transport

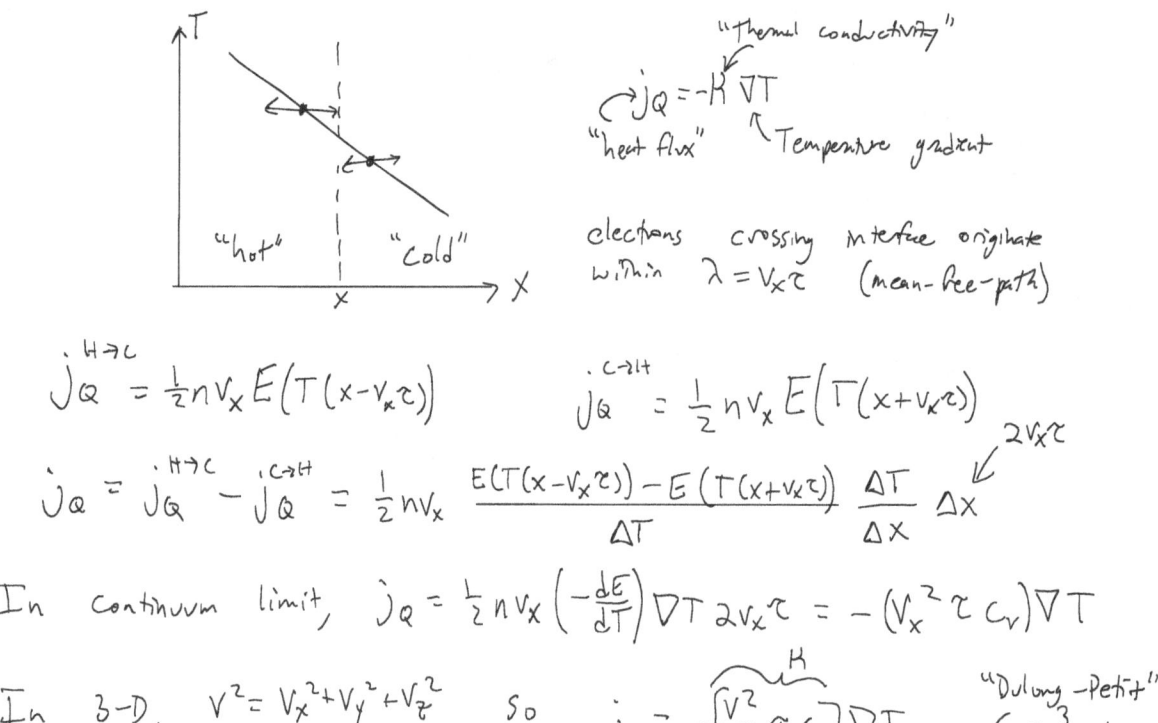

$\vec{j}_Q = -K \vec{\nabla} T$ "Thermal conductivity"
"heat flux" — Temperature gradient

electrons crossing interface originate within $\lambda = v_x \tau$ (mean-free-path)

$$j_Q^{H \to C} = \tfrac{1}{2} n v_x E(T(x - v_x \tau)) \qquad j_Q^{C \to H} = \tfrac{1}{2} n v_x E(T(x + v_x \tau))$$

$$j_Q = j_Q^{H \to C} - j_Q^{C \to H} = \tfrac{1}{2} n v_x \frac{E(T(x-v_x\tau)) - E(T(x+v_x\tau))}{\Delta T} \frac{\Delta T}{\Delta x} \Delta x \quad (2 v_x \tau)$$

In continuum limit, $j_Q = \tfrac{1}{2} n v_x \left(-\frac{dE}{dT}\right) \nabla T \cdot 2 v_x \tau = -(v_x^2 \tau c_v) \nabla T$

In 3-D, $v^2 = v_x^2 + v_y^2 + v_z^2$ so $j_Q = -\left[\overbrace{\frac{v^2}{3} \tau c_v}^{K}\right] \nabla T$

"Dulong–Petit" $c_v = \tfrac{3}{2} n k_B$

Weidemann-Franz Law

Both heat and charge conductivity are proportional to the (undetermined) scattering time τ. Therefore, their ratio can be compared to exp't without any fitting:

$$\frac{K}{\sigma} = \frac{\frac{v^2}{3} \tau c_v}{\frac{n e^2 \tau}{m}} = \frac{2 (\tfrac{1}{2} m v^2)^{\nearrow \tfrac{3}{2} k_B T \text{ by equipartition}} (\tfrac{3}{2} n k_B)}{3 n e^2} = \tfrac{3}{2} \left(\frac{k_B}{e}\right)^2 T = LT$$

"Lorenz number"

$$L = \tfrac{3}{2} \left(\frac{\tfrac{1}{40} eV}{300 K} \cdot \frac{1}{e}\right)^2 \sim 7.5 \times 10^{-9} \frac{V^2}{K^2} \to \left[\frac{(IV)(\tfrac{V}{I})}{K^2}\right] \to \left[\frac{W \Omega}{K^2}\right]$$

Remarkably correct → up to a factor of 2!

But why? even insulators (with no free electrons) have $c_v \sim \tfrac{3}{2} n k_B$ @ RT.
So electron contribution must be negligible!

Thermopower

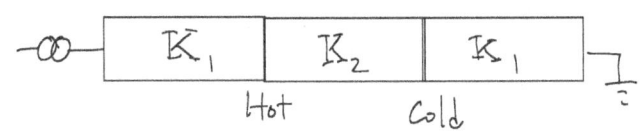

"Seebeck": [diagram: hot → ← cold, electrons drift] electric field "thermopower"
$$\vec{\mathcal{E}} = \overline{K}\nabla T$$
↑ temperature gradient

"Peltier": [diagram: K_1 | K_2 | K_1, Hot, Cold]

heat-driven net velocity

$$V_Q = \frac{1}{2}\frac{[V_x(x-V_x\tau) - V_x(x+V_x\tau)]}{\Delta x}\Delta x \longrightarrow -\tau V_x \frac{dV_x}{dx} = -\frac{\tau}{m}\frac{d(\frac{1}{2}mV_x^2)}{dx}$$

(equipartition)

$$= -\frac{\tau}{m}\frac{d(\frac{1}{2}k_B T)}{dT}\frac{dT}{dx} = -\frac{\tau k_B}{2m}\nabla T$$

Seebeck thermovoltage

In equilibrium, $j = 0$ so $V_Q = -V_{\mathcal{E}} = \frac{e\mathcal{E}}{m}\tau$ ← Drude drift velocity
(heat driven) (electric-field driven)

$$-\frac{\tau k_B}{2m}\nabla T = \frac{e\mathcal{E}}{m}\tau \qquad \text{so} \qquad \mathcal{E} = -\frac{k_B}{2e}\nabla T$$

$$\overline{K} = -\frac{k_B}{2e} = -\frac{1}{2}\left(\frac{\frac{1}{40}eV}{300K}\right) = -40\,\mu V/K \qquad \sim 100\times \text{ too big!}$$

15

Failures of the Drude model

- heat capacity of insulators ($n \to 0$) same as metals @ RT ("Dulong-Petit"). Therefore, it cannot be due solely to kinetic energy of electron gas! Lorenz number calculation correct prediction just a coincidence!

- Thermopower too big by $\sim \times 100$!

- Thermal dependence of conductivity cannot be explained!

- Hall charge carrier sign positive for divalent/trivalent metals!

Sommerfeld model : Quantum mechanics

"noninteracting" "free" electron gas:
$$-\frac{\hbar^2}{2m}\nabla^2 \psi = E\psi \longrightarrow \psi \propto e^{i\vec{k}\cdot\vec{r}} \quad \left(|\vec{k}| = \sqrt{\frac{2mE}{\hbar^2}}\right)$$

Continuous translational symmetry: ψ is eigenstate of $p = \frac{\hbar}{i}\nabla$ so $\hbar\vec{k}$ conserved.

How to fill N states?

At $T=0$:
- Energy minimization ($E \propto k_x^2 + k_y^2 + k_z^2$)
- Pauli exclusion

Result in isotropic state filling $\longrightarrow$

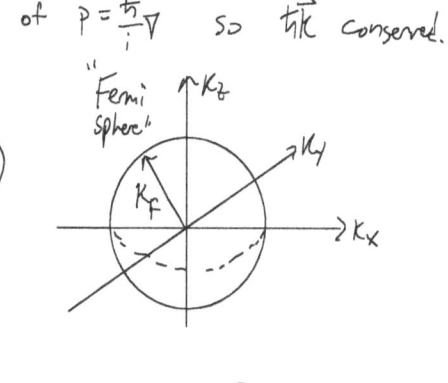

Fermi sphere has volume $\Omega = \frac{4}{3}\pi k_F^3$. How big is k_F?

Density of states

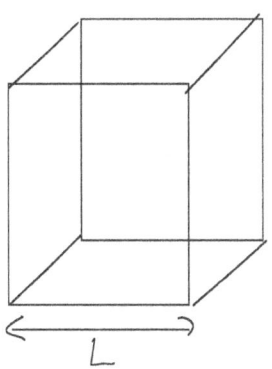

"Free" → no confining potential
          infinite extent

However, electron density finite + fixed

full translational symmetry → arbitrarily periodic

Periodic boundary conditions: $\psi(\vec{r}) = \psi(\vec{r}+L\hat{x}) = \psi(\vec{r}+L\hat{y}) = \psi(\vec{r}+L\hat{z})$

$$e^{i\vec{k}\cdot\vec{r}} = e^{i\vec{k}\cdot(\vec{r}+L\hat{x})} = e^{i\vec{k}\cdot\vec{r}} e^{ik_x L}$$

$$k_x L = 2\pi \cdot \ell$$

Spacing of states along k_x is $\Delta k_x = \frac{2\pi}{L}$

In 3D $\vec{k}$-space, volume $\Delta^3 k = \frac{(2\pi)^3}{\text{Vol.}}$ per state

Fermi sphere has volume $\Omega = \frac{4}{3}\pi k_F^3$ → $N = \overset{\text{spin}}{2} \frac{\frac{4}{3}\pi k_F^3}{(2\pi)^3/\text{Vol}}$ → $n = \frac{k_F^3}{3\pi^2}$

Properties of the Fermi sphere

for "good metal" $n \sim 10^{23} \text{cm}^{-3}$

Radius "Fermi wavevector" $k_F = (3\pi^2 n)^{1/3} \sim 10^8 \text{cm}^{-1}$

On the "Fermi surface" ($|\vec{k}| = k_F$)

"Fermi velocity": $\frac{\hbar k_F}{m} = \frac{6.6\times10^{-16}\text{eV s} \cdot 9\times10^{20}\frac{\text{cm}^2}{\text{s}^2}}{5\times10^5 \text{eV}} \cdot 10^8 \text{cm}^{-1} \sim 10^8 \frac{\text{cm}}{\text{s}}$

"Fermi energy": $E_F = \frac{\hbar^2 k_F^2}{2m} = \frac{(6.6\times10^{-16}\text{eV s})^2}{2.5\times10^5 \text{eV}} 9\times10^{20}\frac{\text{cm}^2}{\text{s}^2} \cdot (10^8\text{cm}^{-1})^2 \sim 1\text{-}10\text{eV}$

"Fermi Temperature": $T_F = \frac{E_F}{k_B} = \frac{1\text{-}10\text{eV}}{\frac{1}{40}\text{eV}/300\text{K}} \sim 10^4 \text{-} 10^5 \text{K} \gg RT$

So $T=0$ approximation (abrupt boundary between filled and unfilled states at E_F) is expected to capture even RT behavior!

17

Finite temperature

Zero-temp approx works for $n \sim 10^{23} \text{cm}^{-3}$ metal, but will fail for sufficiently dilute density where

$$E_F \sim k_B T = \frac{\hbar^2}{2m}(3\pi^2 n)^{2/3}$$

Then, (Note: convenient to remember $hc \approx 1.24 \, eV \cdot \mu m$)

$$n = \frac{\left(\frac{2mk_BT}{\hbar^2}\right)^{3/2}}{3\pi^2} \sim \frac{\left(\frac{2 \cdot 5\times 10^5 eV \cdot 1/40 \, eV}{\hbar^2 c^2}\right)^{3/2}}{3\pi^2} \sim \frac{\left(\frac{(2\pi)^2 \cdot \frac{5}{20} \times 10^5 eV}{(10^{-4} eV \, cm)^2}\right)^{3/2}}{3\pi^2} \sim 10^{19} \text{cm}^{-3}$$

This is a characteristic density above which Pauli exclusion is dominant (a "degenerate" gas). For lower densities ("nondegenerate") we need to account for electron excitation above E_F.

Total energy

Sum over contributions from every electron in Fermi sphere:

$$E = 2 \sum_{|\vec{k}|<k_F} \frac{\hbar^2 k^2}{2m}$$

Since $\Delta^3 k = \frac{(2\pi)^3}{\text{Vol.}}$,

$$E = \frac{2 \cdot \text{Vol.}}{(2\pi)^3} \sum_{|\vec{k}|<k_F} \frac{\hbar^2 k^2}{2m} \Delta^3 k \longrightarrow \frac{2 \cdot \text{Vol.}}{(2\pi)^3} \int_{\text{Fermi sphere}} \frac{\hbar^2 k^2}{2m} d^3k = \frac{2 \cdot \text{Vol.}}{(2\pi)^3} \int \frac{\hbar^2 k^2}{2m} 4\pi k^2 dk$$

Energy density $u = \frac{E}{\text{vol}} = \frac{\hbar^2}{2\pi^2 m} \frac{k^5}{5}\bigg|_0^{k_F} = \frac{\hbar^2 k_F^5}{10\pi^2 m}$

Avg energy per electron $\frac{u}{n} = \frac{\hbar^2 k_F^5 / 10\pi^2 m}{k_F^3 / 3\pi^2} = \frac{3}{10} \frac{\hbar^2 k_F^2}{m} = \frac{3}{5} E_F$

(Larger than $\frac{E_F}{2}$ since the density of states increases with E)

Fermi pressure

$$P = -\left.\frac{\partial E}{\partial V}\right|_N = -\frac{\partial}{\partial V}\left(\frac{3}{5}E_F N\right)\bigg|_N = -\frac{3}{5}N\frac{\partial}{\partial V}\left(\frac{\hbar^2}{2m}\left(3\pi^2\frac{N}{V}\right)^{2/3}\right)\bigg|_N$$

$$= -\frac{3}{5}N\frac{\hbar^2}{2m}\frac{\left(3\pi^2\frac{N}{V}\right)^{2/3}}{V}\left(-\frac{2}{3}\right) = \frac{2}{5}E_F n$$

Assuming $n \sim 10^{23}\,cm^{-3}$, $P \sim 10^{(23-24)}\,\frac{eV}{cm^3}$

Convert to more familiar units:

$$10^{24}\frac{eV}{cm^3} \cdot \frac{1.6\times10^{-19}\,J}{eV} \cdot \frac{10^6\,cm^3}{m^3} \simeq 10^{11}\,\frac{N}{m^2} \sim 10^6\,atm\,!$$

"Compressivity" $= \frac{-\frac{dV}{dP}}{V} \longrightarrow$ "bulk modulus" $B = \frac{1}{compressivity} = -V\frac{dP}{dV}$

Since $P \propto V^{-5/3}$, $B = -V\left(-\frac{5}{3}\frac{P}{V}\right) = \frac{5}{3}P = \frac{2}{3}E_F n \sim 10^6\,atm\,!$

So Pauli exclusion contributes to the incompressibility of metals!

Electrical Conductivity

Each plane-wave electron has a different momentum, so eqn. of motion useless!

$f(\vec{k})$ occupation function of filled states (in equilibrium $|\vec{\mathcal{E}}|=0$, $f_0(\vec{k})=1$ for $|\vec{k}|<k_F$)

Current density

$$j = -2e\sum_{\substack{filled \\ Volume}} V_k \quad \xrightarrow{d^3k = \frac{(2\pi)^3}{Volume}} \quad j = -2e\int \underbrace{f_0(\vec{k})}_{even}\underbrace{v_z(\vec{k})}_{odd}\frac{d^3k}{(2\pi)^3} = 0 \quad \text{in equilibrium } (|\vec{\mathcal{E}}|=0)$$

for $\mathcal{E}\neq 0$, what is $f(\vec{k}) = f_0(\vec{k}) + g(\vec{k})$?

"Boltzmann transport equation": "excitation" = "relaxation"

"excitation": $\dfrac{df}{dt} = \dfrac{df}{dp_z}\dfrac{dp_z}{dt} = \dfrac{df}{dk_z}\dfrac{-e\mathcal{E}_z}{\hbar}$ $\quad\Rightarrow\quad$ algebraic solution $g(k) = \dfrac{e\mathcal{E}_z \tau}{\hbar}\dfrac{df}{dk_z}$

"relaxation": $\dfrac{df}{dt} = -\left(\dfrac{f(\vec{k})-f_0(\vec{k})}{\tau}\right) = -\dfrac{g(\vec{k})}{\tau}$

"relaxation time approximation"

Interpretation

$$f(\vec{k}) = f_0(\vec{k}) + \frac{eE_z\tau}{\hbar}\frac{df}{dk_z} \xrightarrow{\text{taylor expansion}} f_0\left(\vec{k} + \frac{eE_z\tau}{\hbar}\hat{z}\right)$$

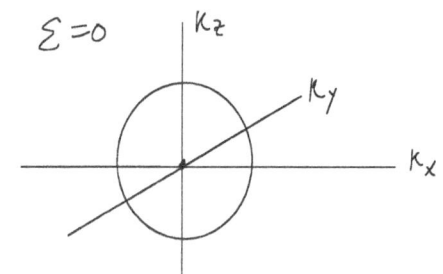
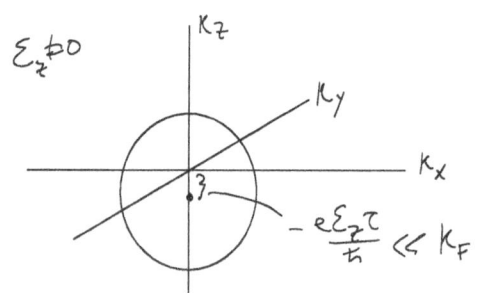

So electric field creates asymmetry by rigidly shifting Fermi sphere!

Current density: $j = -2e\int g(\vec{k})\frac{\hbar k_z}{m}\frac{d^3k}{(2\pi)^3} = -2e\int \frac{eE_z\tau}{\hbar}\frac{df}{dk_z}\frac{\hbar k_z}{m}\frac{d^3k}{(2\pi)^3}$

write $\frac{df}{dk_z} = \frac{df}{dk}\frac{dk}{dk_z}$ where $\frac{dk}{dk_z} = \frac{d}{dk_z}\sqrt{k_x^2+k_y^2+k_z^2} = \frac{\frac{1}{2}\cdot 2k_z}{\sqrt{k_x^2+k_y^2+k_z^2}} = \frac{k_z}{k}$

$$j = \left[2\frac{e^2\tau}{m}\int\left(-\frac{df}{dk}\right)\frac{k_z^2}{k}\frac{d^3k}{(2\pi)^3}\right]E_z$$

Fermi Surface Integration

For quasi-spherical $f(\vec{k})$, $-\frac{df}{dk} = \delta(|\vec{k}|-k_F)$
This converts our <u>volume</u> integral into a surface integral!

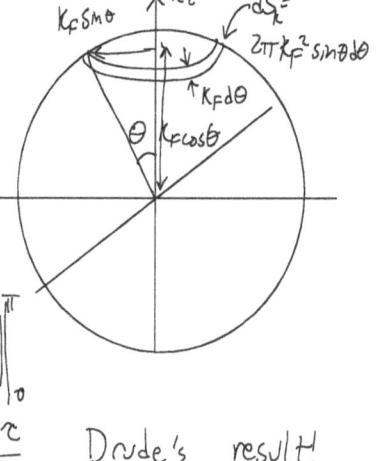

$\sigma = \frac{1}{4\pi^3}\frac{e^2\tau}{m}\int_{\text{Fermi surface}} \frac{k_z^2}{k}dS_k$. In spherical coords,

$= \frac{1}{4\pi^3}\frac{e^2\tau}{m}\int \frac{(k_F\cos\theta)^2}{k_F}2\pi k_F^2 \sin\theta\, d\theta = \frac{k_F^3}{2\pi^2}\frac{e^2\tau}{m}\left[-\frac{\cos^3\theta}{3}\right]_0^\pi$ $\overbrace{}^{2/3}$

$= \frac{k_F^3}{3\pi^2}\frac{e^2\tau}{m} = \frac{ne^2\tau}{m}$ Drude's result!

→ Fermi Surface is <u>fundamental</u> in determining transport properties because only there can electrons make infinitesimal transitions to unfilled states

→ total density n appears only because of integration over sphere, but unlike Drude, not all electrons contribute!

N.B.: Fermi Surface is <u>not</u> perfectly spherical in real metals due to lattice potential, etc.

"Fluctuation-Dissipation Theorem"

Rewrite our integral as Fermi surface area times avg. value of integrand:

$$\sigma = \frac{1}{4\pi^3}\frac{e^2}{m}\frac{\langle k_z^2 \tau\rangle}{k_F} 4\pi k_F^2 = \frac{e^2 k_F}{\pi^2 m}\frac{m^2}{\hbar^2}\langle \left(\frac{\hbar k_z}{m}\right)^2 \tau\rangle \overset{\text{in 3D}}{=} \frac{e^2 k_F m}{3\pi^2 \hbar^2}\langle v^2 \tau\rangle$$

$$\sigma = ne\mu = e\mu \frac{k_F^3}{3\pi^2}, \quad \text{so:}$$

$$\frac{e^2 k_F m}{3\pi^2 \hbar^2}\langle v^2 \tau\rangle = e\mu \frac{k_F^3}{3\pi^2} \longrightarrow \mu = \frac{2m}{\hbar^2 k_F^2}\frac{3}{2}\langle \frac{v^2 \tau}{3}\rangle e$$

$\langle \frac{\lambda^2}{3\tau}\rangle$ Variance of spatial fluctuation in time τ "Diffusion Coef" D

$$\frac{D}{\mu} = \frac{2}{3}\frac{E_F}{e} \quad \text{(degenerate regime)}$$

So, "fluctuations" at equilibrium (Diffusion coef) are related to "dissipation" due to relaxation toward equilibrium (mobility)!

In nondegenerate regime,

$$\frac{D}{\mu} = \frac{\langle \frac{v^2 \tau}{3}\rangle}{\frac{e\tau}{m}} = \frac{\frac{2}{3m}\langle \frac{1}{2}mv^2 \tau\rangle}{\frac{e\tau}{m}} \xrightarrow{\text{equipartition}} \frac{\frac{2}{3m}\left(\frac{3}{2}k_B T\right)\tau}{\frac{e\tau}{m}} = \frac{k_B T}{e}$$

"Einstein relation" Derived in context of Brownian motion

Beyond Ohm's law: Boundaries and

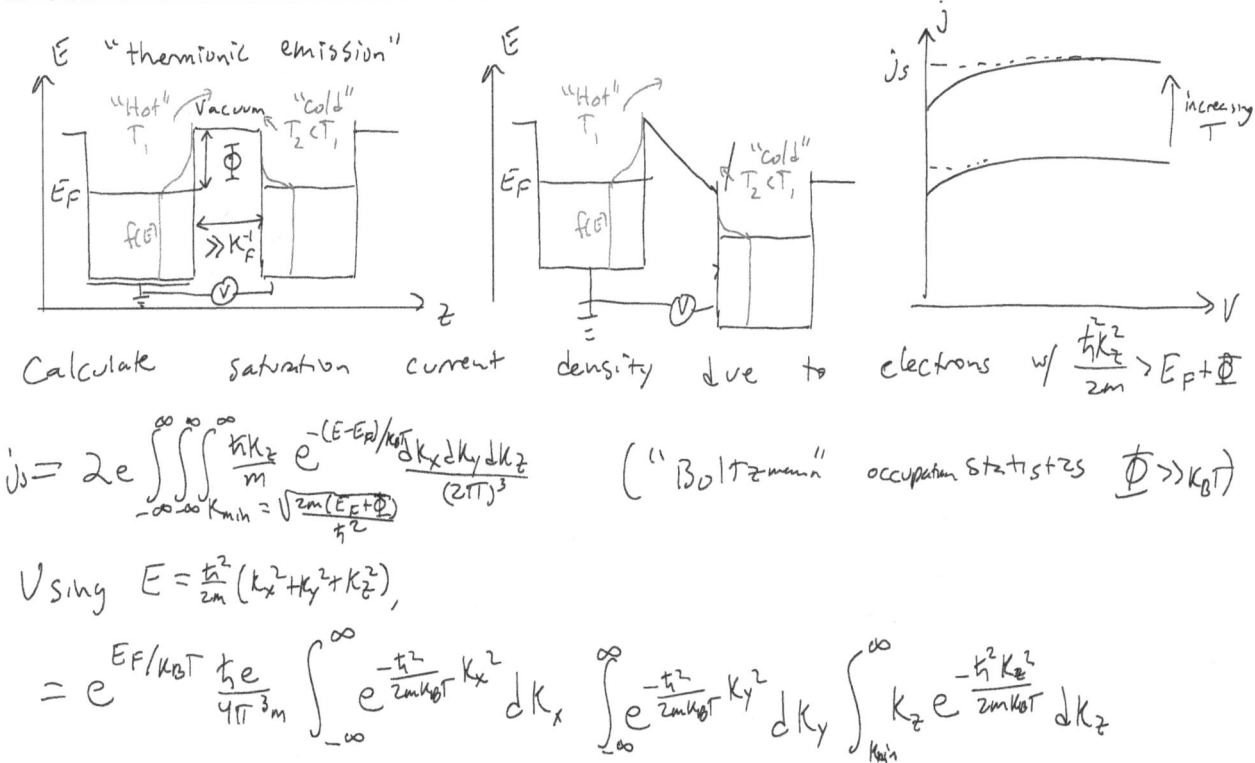

Calculate saturation current density due to electrons w/ $\frac{\hbar^2 k_z^2}{2m} > E_F + \Phi$

$$j_s = 2e \iiint_{-\infty\quad -\infty\quad k_{min} = \sqrt{\frac{2m(E_F+\Phi)}{\hbar^2}}}^{\infty\quad\infty\quad\infty} \frac{\hbar k_z}{m}\frac{e^{-(E-E_F)/k_B T}}{(2\pi)^3} dk_x dk_y dk_z \quad (\text{"Boltzmann" occupation statistics } \underline{\Phi \gg k_B T})$$

Using $E = \frac{\hbar^2}{2m}(k_x^2+k_y^2+k_z^2)$,

$$= e^{E_F/k_B T}\frac{\hbar e}{4\pi^3 m}\int_{-\infty}^{\infty} e^{-\frac{\hbar^2}{2mk_B T}k_x^2}dk_x \int_{-\infty}^{\infty} e^{-\frac{\hbar^2}{2mk_B T}k_y^2}dk_y \int_{k_{min}}^{\infty} k_z e^{-\frac{\hbar^2 k_z^2}{2mk_B T}}dk_z$$

Integral evaluation

Since gaussian integral $\int_{-\infty}^{\infty} e^{-ax^2} dx = \sqrt{\frac{\pi}{a}}$

$$j_s = e^{E_F/k_BT} \frac{\hbar e}{4\pi^3 m} \left(\sqrt{\frac{2\pi m k_B T}{\hbar^2}}\right)^2 \int_{K_{min}^2}^{\infty} e^{-\frac{\hbar^2}{2mk_BT}k_z^2} k_z^2 \frac{d(k_z^2)}{2} \qquad (X \equiv k_z^2)$$

$$= e^{E_F/k_BT} \frac{\hbar e}{4\pi^3 m} \frac{2\pi m k_B T}{\hbar^2} \frac{1}{2}\left(-\frac{2mk_BT}{\hbar^2} e^{-\frac{\hbar^2}{2mk_BT}X}\right)\bigg|_{X=K_{min}^2 = \frac{2m(E_F+\Phi)}{\hbar^2}}^{\infty}$$

$$= \frac{ek_BT}{2\pi^2 \hbar} \frac{mk_BT}{\hbar^2} e^{-\Phi/k_BT} = \underbrace{\frac{4\pi m k_B^2 e}{h^3}}_{\text{"Richardson Coef" } (A_0)} T^2 e^{-\Phi/k_BT} \quad \text{"Richardson–Dushman"}$$
"Shottky" saturation current density

$$A_0 \approx 4\pi \frac{5\times 10^5 eV/c^2 \left(\frac{1}{40}\frac{eV}{300K}\right)^2 e}{h^3} \approx \frac{4\times 10^{-2} \frac{eV^3}{K^2} e}{(hc)^2 h} = \frac{4\times 10^{-2} \frac{eV^3}{K^2} \cdot 1.6\times 10^{-19} C}{(1.24\times 10^{-4} cm\, eV)^2 \cdot 4\times 10^{-15} eV\cdot s} \cong 120 \frac{A}{cm^2 K^2}$$

Elastic, coherent scattering

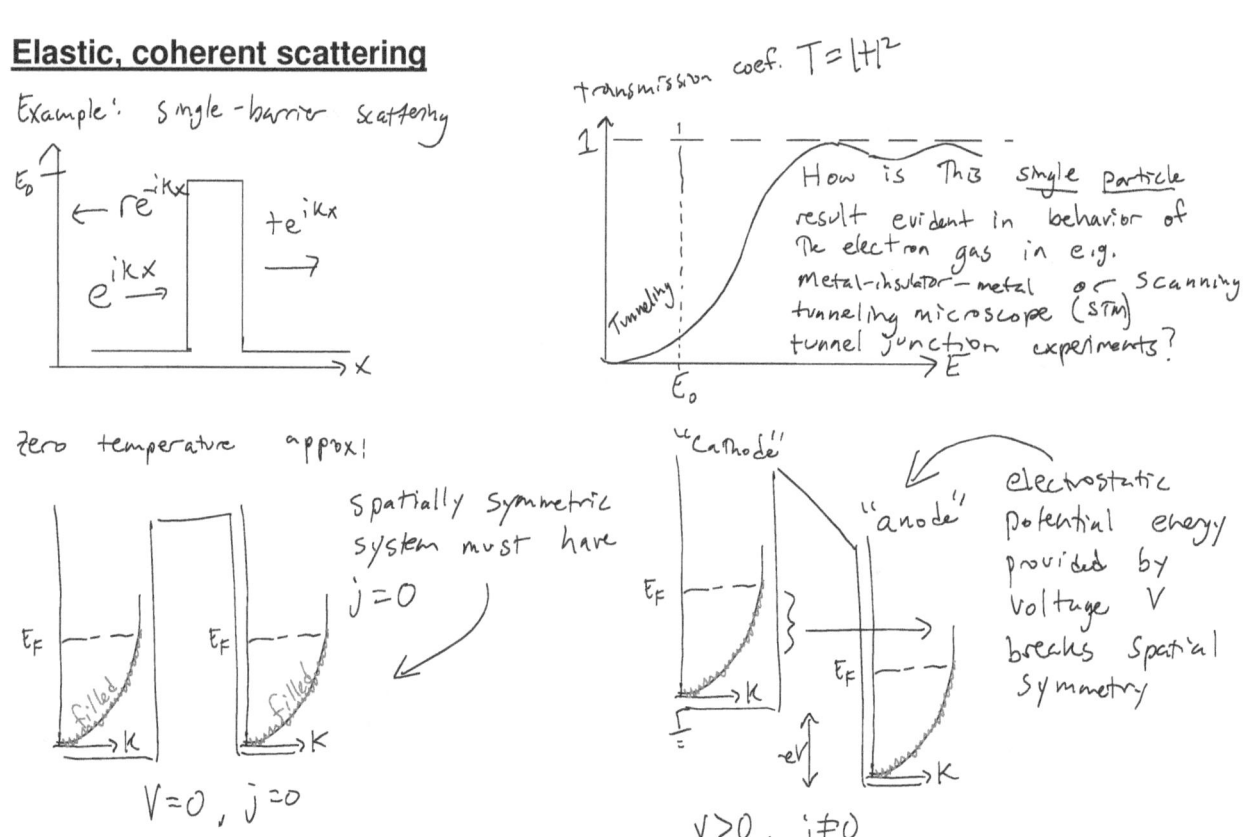

Summation over contributing states

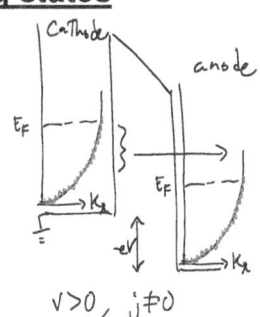

$V > 0, \; j \neq 0$

Energy conservation:

$$\underbrace{\frac{\hbar^2 k_x^2}{2m}}_{(\text{kinetic})} + \underbrace{eV}_{\text{cathode (potential)}} > \underbrace{E_F = \frac{\hbar^2 k_F^2}{2m}}_{\text{anode}}$$

$$k_x > k_{min} = \sqrt{\frac{2m}{\hbar^2}(E_F - eV)} = \sqrt{k_F^2 - \frac{2meV}{\hbar^2}}$$

1-D current: $\quad I(V) = 2e \int_{k_{min}(V)}^{k_F} \frac{\hbar k_x}{m} \, T(E(k_x), V) \, \frac{dk_x}{2\pi}$

← transmission coef
← solve numerically.

Numerical integration

See, e.g. STM, M-I-M junctions, etc!

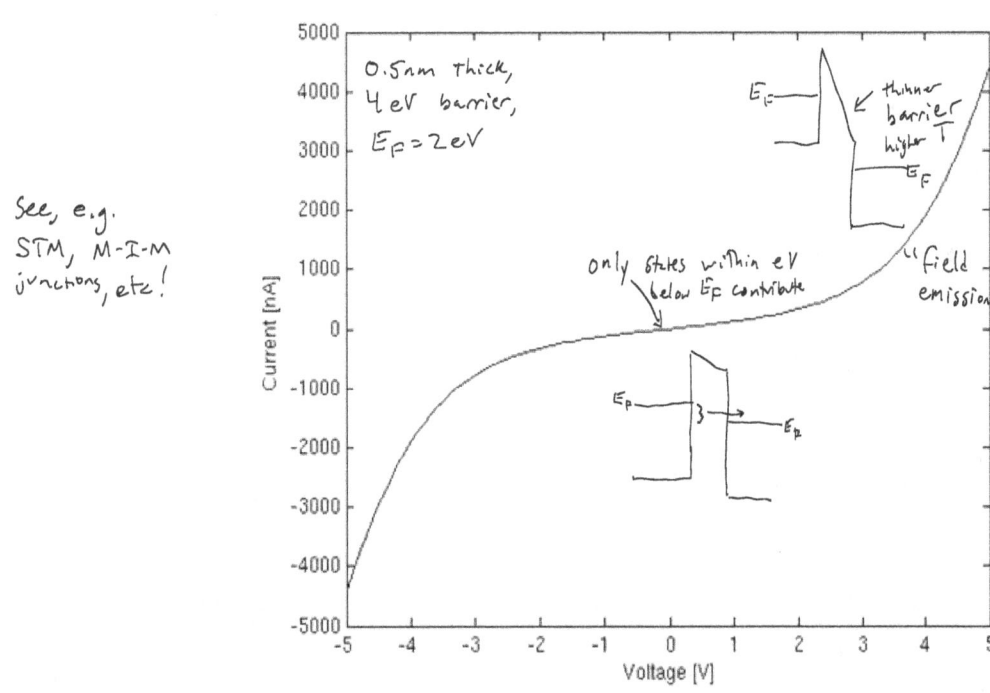

0.5 nm thick, 4 eV barrier, $E_F = 2 eV$

thinner barrier higher T

only states within eV below E_F contribute

"field emission"

23

Double-barrier resonant scattering

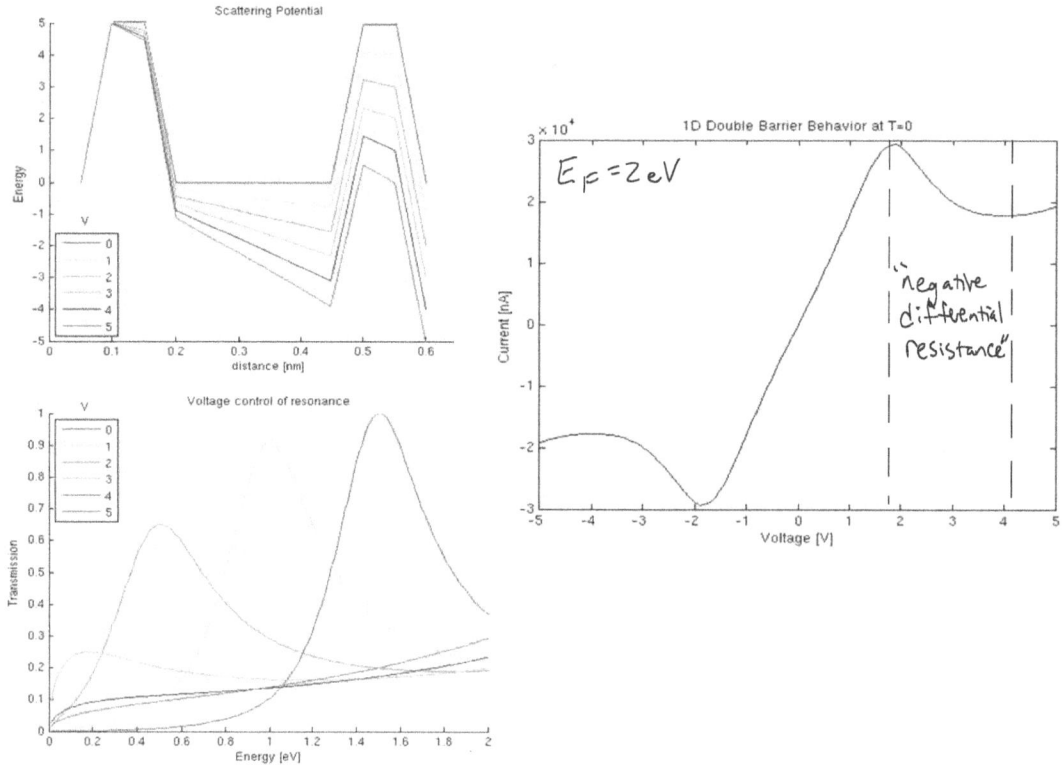

$E_F = 2 eV$

"negative differential resistance"

Resonant tunneling through quantum wells at frequencies up to 2.5 THz

T. C. L. G. Sollner, W. D. Goodhue, P. E. Tannenwald, C. D. Parker, and D. D. Peck
Lincoln Laboratory, Massachusetts Institute of Technology, Lexington, Massachusetts 02173

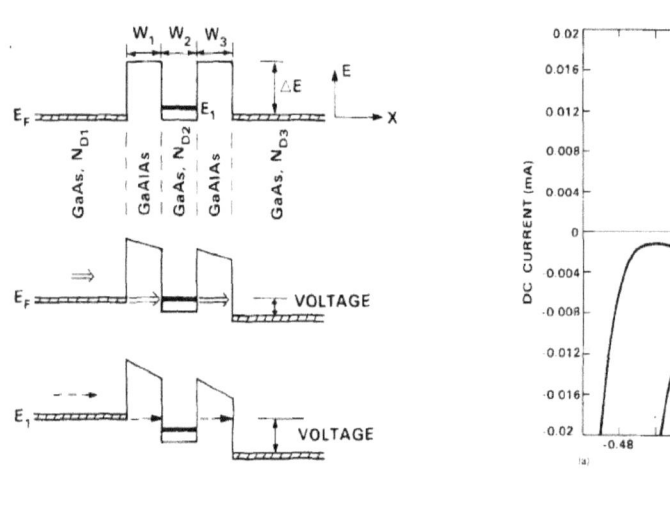

NDR is important to "cancel" dissipation in LCR oscillators!

Extension to 2D

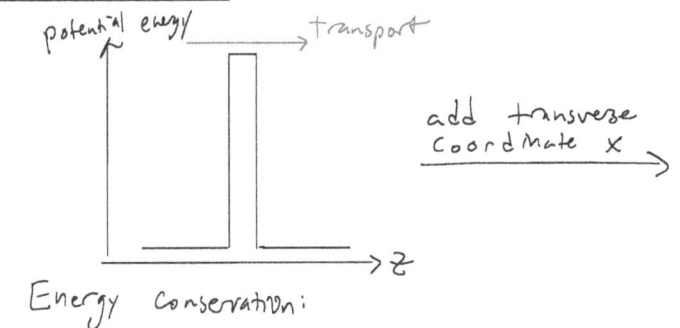
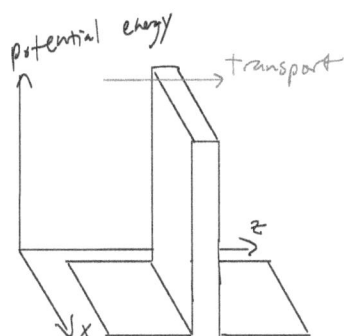

Energy Conservation:
(cathode)   (anode)

$$E + eV > E_F$$

$$\frac{\hbar^2}{2m}(K_x^2 + K_z^2) > \frac{\hbar^2 K_F^2}{2m} - eV$$

$$K_x^2 + K_z^2 > K_{min}^2 = \left(\sqrt{K_F^2 - \frac{2meV}{\hbar^2}}\right)^2 \quad \text{(a disk!)}$$

So current density

$$J = 2e \iint_{K_{min} < |\vec{K}| < K_F} \frac{\hbar K_z}{m} T(K_z, V) \frac{dK_x}{2\pi} \frac{dK_z}{2\pi}$$

Integral evaluation

The integrand only depends on K_z so if we choose our area infinitesimal wisely, we can simplify this 2D integral into 1-D!

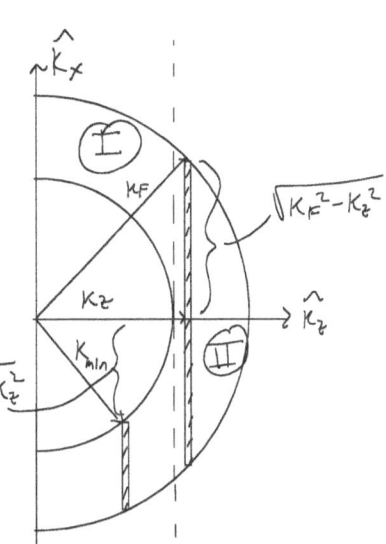

$$J_{2D} = 2e \left[\int_0^{K_{min}} \overset{\text{I}}{\frac{\hbar K_z}{m} T(K_z, V) \, 2\left(\sqrt{K_F^2 - K_z^2} - \sqrt{K_{min}^2 - K_z^2}\right) \frac{dK_z}{(2\pi)^2}} \right.$$

$$\left. + \int_{K_{min}}^{K_F} \overset{\text{II}}{\frac{\hbar K_z}{m} T(K_z, V) \, 2\sqrt{K_F^2 - K_z^2} \frac{dK_z}{(2\pi)^2}} \right]$$

Extension to 3D

Now, the contributing states satisfying $k_{min} < |\vec{k}| < k_F$ and $k_z > 0$ fill a hemispherical shell, and our infinitesimal becomes a disk of thickness dk_z, and radius determined exactly by the lengths described above in 2D.

① for $0 < k_z < k_{min}$: $dV = \pi\left[(k_F^2 - k_z^2) - \left(k_F^2 - \frac{2meV}{\hbar^2} - k_z^2\right)\right] dk_z$

$\qquad = \pi \frac{2meV}{\hbar^2} dk_z$

② for $k_{min} < k_{min} < k_F$: $dV = \pi(k_F^2 - k_z^2) dk_z$

Linear response in 1D

For small V, only states close to E_F contribute. Further, barrier potential is only negligibly deformed. Therefore, approximate $T(E(k), V)$ as a constant + remove from integral

$$I = 2e \int_{k_{min}}^{k_F} \frac{\hbar k}{m} T(E(k), V) \frac{dk}{2\pi} \sim 2e \frac{\hbar}{2\pi m} T(E_F) \int_{k_{min}}^{k_F} k\, dk$$

$$= \frac{e\hbar}{\pi m} T \left.\frac{k^2}{2}\right|_{k_{min} = \sqrt{k_F^2 - \frac{2meV}{\hbar^2}}}^{k_F}$$

$$= \frac{e\hbar}{2\pi m} T \left(k_F^2 - \left(k_F^2 - \frac{2meV}{\hbar^2}\right)\right) = \frac{e^2}{\pi \hbar} TV = \left(\frac{2e^2}{h} T\right) V$$

Note that $T \leq 1$ even for no scatterer! 1-D current is bounded by the "quantum of conductance" $\frac{2e^2}{h}$:

$$\frac{2e^2}{h} = \frac{e^2}{\pi \hbar} = \frac{e^2}{\pi \cdot 6.6 \times 10^{-16} eV\cdot s} = \frac{1.6 \times 10^{-19} A}{\pi \cdot 6.6 \times 10^{-19} V} \approx \frac{1}{13 k\Omega} \quad \left(\frac{h}{e^2} \sim 26 k\Omega \text{ "von Klitzing"}\right)$$

Experiment: Quantum Point Contact

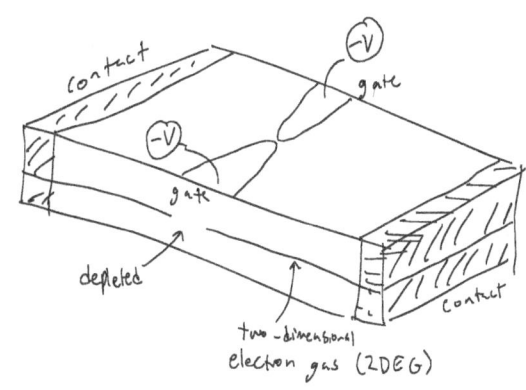

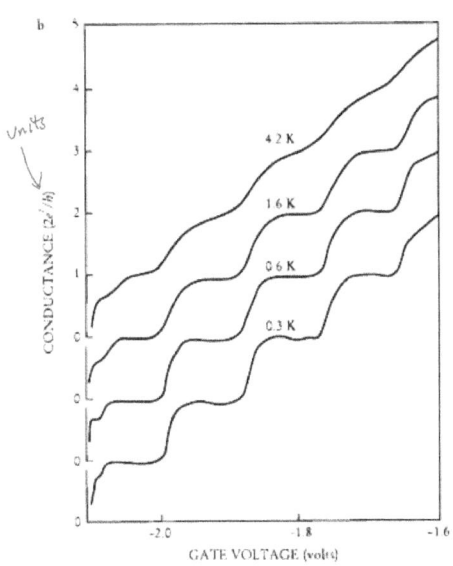

Each 1-D channel contributes $\frac{2e^2}{h}$, so as the constriction gets wider with increasing gate voltage, more channels can fit and we see a conductance staircase
→ width of 1-D conductor set by Fermi wavelength $\left(\begin{array}{l}\lambda_F \sim 50\,nm \\ \text{for } n \sim 10^{11}\,cm^{-2}\end{array}\right)$

Optical "Quantized Conduction"

This phenomenon can also be seen in optical transmission of diffuse monochromatic photons thru a slit (where $\lambda_F \to \lambda$)

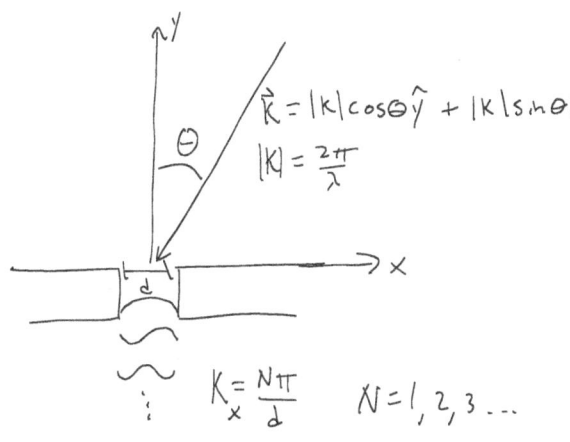

$\vec{K} = |K|\cos\theta\,\hat{y} + |K|\sin\theta\,\hat{x}$
$|K| = \frac{2\pi}{\lambda}$

$K_x = \frac{N\pi}{d}$ $N = 1, 2, 3 \ldots$

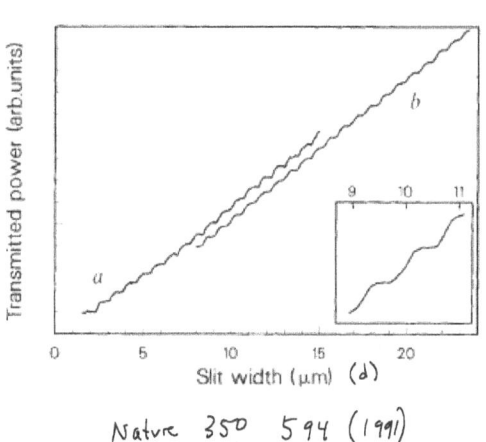

Nature **350** 594 (1991)

Microscopic perspective on 1-D transport

[Diagram: Energy vs x showing 1-D leads on left and right with a 0-D region in the middle containing two levels. Left lead has occupation f_1 with coupling γ_1 and currents I_1 flowing into region with occupation N; right lead has f_2 with coupling γ_2 and current I_2. Coupling strength in [eV].]

In steady state, $I_1 = -I_2$

$$e\frac{\gamma_1}{\hbar}\underbrace{[f_1(1-N) - N(1-f_1)]}_{\text{forward} \quad \text{backward}} = -e\frac{\gamma_2}{\hbar}[f_2(1-N) - N(1-f_2)]$$

$$\gamma_1(f_1 - N) = -\gamma_2(f_2 - N)$$

$$N(\gamma_1 + \gamma_2) = \gamma_1 f_1 + \gamma_2 f_2$$

$$N = \frac{\gamma_1 f_1 + \gamma_2 f_2}{\gamma_1 + \gamma_2}$$

A Contradiction?

Total current flowing from left lead to right:

$$I = I_1 = e\frac{\gamma_1}{\hbar}(f_1 - N) = e\frac{\gamma_1}{\hbar}\left(f_1 - \frac{\gamma_1 f_1 + \gamma_2 f_2}{\gamma_1 + \gamma_2}\right)$$

$$= e\frac{\gamma_1}{\hbar}\left(\frac{f_1(\gamma_1 + \gamma_2) - \gamma_1 f_1 - \gamma_2 f_2}{\gamma_1 + \gamma_2}\right)$$

$$= \frac{e}{\hbar}\frac{\gamma_1 \gamma_2}{\gamma_1 + \gamma_2}(f_1 - f_2)$$

In strong coupling regime, $\gamma_1, \gamma_2 \to \infty$, then $I \to \infty$ even if only one state is between Fermi Energy of the leads at vanishing applied voltage! This prediction directly contradicts the (experimentally verified) expected finite quantum of conduction. How to reconcile?

Uncertainty Principle

$\Delta E \Delta t \sim \hbar$ → Consequence of Energy/time Fourier transform pair.

Trivial Example: "Bound state" eigenfunctions of Schrödinger eqn

$$i\hbar \frac{d}{dt}\Psi = H\Psi$$

time-independent H → Ψ is separable: $\Psi(x,t) = \varphi(x)\phi(t)$, $\phi(t) = e^{-i\frac{E_n}{\hbar}t}$

Uncertainty in time $\Delta t \to \infty$ since $\phi^*\phi = 1$ for all time

Fourier transform $\int_{-\infty}^{\infty} e^{i(\omega - \frac{E_n}{\hbar})t} dt = \delta(\omega - \frac{E_n}{\hbar}) = \delta(E - E_n)$ width → 0

So energy + time uncertainty $\begin{array}{c}\Delta t \to \infty \\ \Delta E \to 0\end{array}$ such that $\Delta E \Delta t \sim \hbar$

→ we have a <u>discrete</u> spectrum of bound states!

State Broadening: DOS

for a "conductor" state coupled to "leads", time dependence

$$\phi(t) = e^{-i\frac{E_n}{\hbar}t} e^{-\frac{|t|}{\tau}}, \text{ where } \frac{1}{\tau} = \frac{\gamma_1 + \gamma_2}{\hbar} \text{ is tunneling rate from 1 lead to other}$$

Fourier transform

$$\int_{-\infty}^{\infty} e^{-i\frac{E_0}{\hbar}t} e^{-\frac{|t|}{\tau}} e^{i\omega t} dt = \int_{-\infty}^{0} e^{i(\omega - \frac{E_0}{\hbar} - \frac{i}{\tau})t} dt + \int_{0}^{\infty} e^{i(\omega - \frac{E_0}{\hbar} + \frac{i}{\tau})t} dt$$

$$= \frac{e^{i(\omega - \frac{E_0}{\hbar} - \frac{i}{\tau})t}}{i(\omega - \frac{E_0}{\hbar} - \frac{i}{\tau})}\bigg|_{-\infty}^{0} + \frac{e^{i(\omega - \frac{E_0}{\hbar} + \frac{i}{\tau})t}}{i(\omega - \frac{E_0}{\hbar} + \frac{i}{\tau})}\bigg|_{0}^{\infty} = \frac{1}{i(\omega - \frac{E_0}{\hbar} - \frac{i}{\tau})} - \frac{1}{i(\omega - \frac{E_0}{\hbar} + \frac{i}{\tau})}$$

$$= 2\frac{\frac{1}{\tau}}{(\omega - \frac{E_0}{\hbar})^2 + (\frac{1}{\tau})^2} \xrightarrow{\text{Normalization}} \int_{-\infty}^{\infty} \frac{\frac{2}{\pi}\frac{1}{\tau}}{(\omega - \frac{E_0}{\hbar})^2 + (\frac{1}{\tau})^2} d\omega = 1$$

So we have smeared our discrete state $\delta(E - E_n)$ into a "Density of States" $D(E) = \frac{\frac{2}{\pi}(\gamma_1 + \gamma_2)}{(E - E_0)^2 + (\gamma_1 + \gamma_2)^2}$ (Lorentzian in E)

State Broadening: Transmission coefficient

You have seen this before!: increasing coupling w/ thinner barriers in resonant 1D scatterers

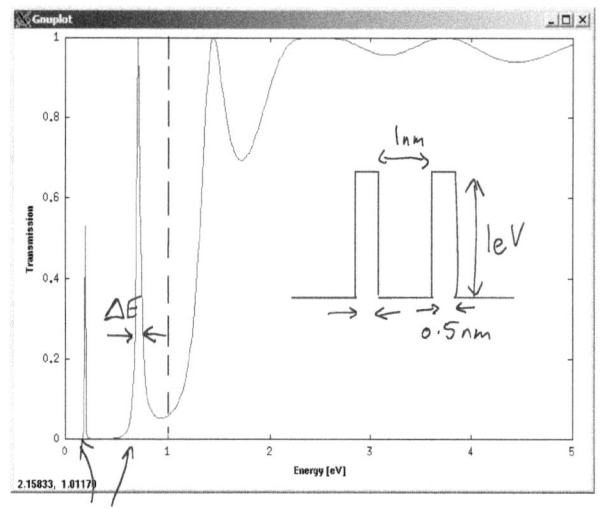 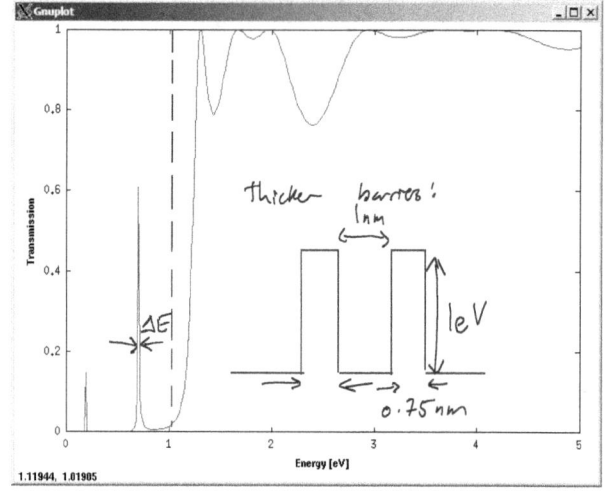

resonances due to quasi-bound states: since $\Delta E \Delta t \sim \hbar$ (frequency-time uncertainty rel'n), the resonances get wider as the quasi-bound state is tunnel-coupled stronger to the continuum of scattering states.

Sum over DOS

$$I = \frac{e}{\hbar} \int_{-\infty}^{\infty} \frac{\gamma_1 \gamma_2}{\gamma_1 + \gamma_2} (f_1 - f_2) D(E) dE$$

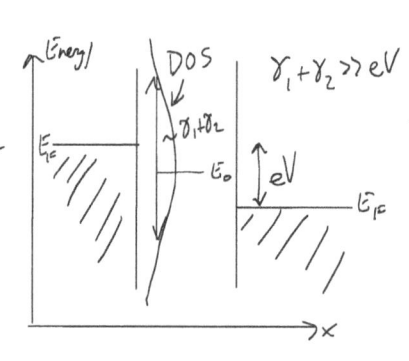 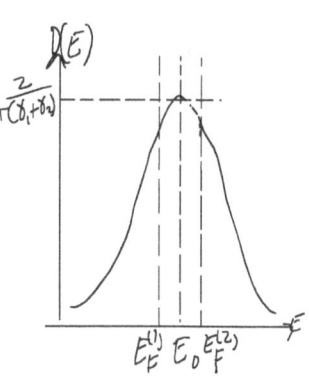

@ $T=0$, and no energy dep. in γ's:

$$I = \frac{e}{\hbar} \frac{\gamma_1 \gamma_2}{\gamma_1 + \gamma_2} \int_{E_F - eV}^{E_F} D(E) dE \xrightarrow{\gamma_1 + \gamma_2 \gg eV} I \simeq \frac{e}{\hbar} \frac{\gamma_1 \gamma_2}{\gamma_1 + \gamma_2} \frac{2/\pi}{\gamma_1 + \gamma_2} eV$$

for $\gamma_1 = \gamma_2$, $I = \frac{e^2}{\hbar \cdot 2\pi} V = \frac{e^2}{h} V$

quantum of conductance (per spin channel)

So conductance quantization is maintained!

Paradox averted, thanks to uncertainty relation!

Electrostatics

Where does the conductor state lie?

Poisson's equation $\nabla \epsilon \nabla V = \rho$

$$\frac{\epsilon_1 \frac{V}{\ell_1} - \epsilon_2 \frac{(V_D - V)}{\ell_2}}{(\ell_1 + \ell_2)/2} = \frac{\Delta N e}{(\ell_1 + \ell_2)/2} \longrightarrow C_S V - C_D (V_D - V) = \Delta N e$$

Solve for potential energy: $eV = \frac{C_D}{C_S + C_D} eV_D + \underbrace{\frac{e^2}{C_S + C_D} \Delta N}_{\text{Single electron charging energy}}$

If $\Delta E > k_B T$ and broadening γ_1, γ_2, we have **no** current at finite bias: "Coulomb blockade."

Coulomb Blockade

Approximate a conductor as a sphere $C = \frac{q}{V} = \frac{q}{\frac{q}{4\pi\epsilon_0 \epsilon_r R}} = 4\pi\epsilon_0 \epsilon_r R$

Single-electron charging energy $\frac{e^2}{C} = \frac{e^2}{4\pi\epsilon_0 \epsilon_r R}$

Since $\frac{e^2}{4\pi\epsilon_0 \hbar c} = \alpha$, $\frac{e^2}{4\pi\epsilon_0} = \alpha \hbar c = \frac{1}{137} \cdot 6.6 \times 10^{-16} \, eV \cdot s \cdot 3 \times 10^{10} \, cm/s$

$\sim 1.4 \times 10^{-7} \, cm \cdot eV$

If $\frac{e^2}{4\pi\epsilon_0 \epsilon_r R} > k_B T \simeq \frac{1}{40} eV @ 300K \, (RT)$, we expect to see Coulomb blockade.

$R < \frac{1.4 \times 10^{-7} \, eV \cdot cm}{10} \cdot \frac{40}{eV} \sim 5 \times 10^{-7} \, cm \, (5 \, nm)$

$\epsilon_r \nearrow$

lower temps allow larger length scales ....

Coulomb Diamond

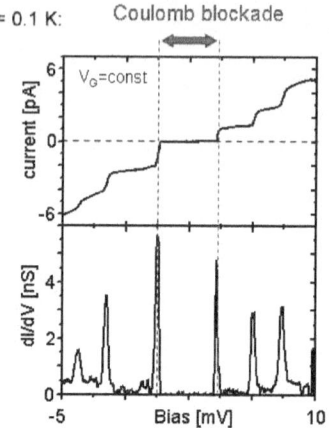

with gates (here, two: g1 and g2)

Conductance plot: "Coulomb diamond"

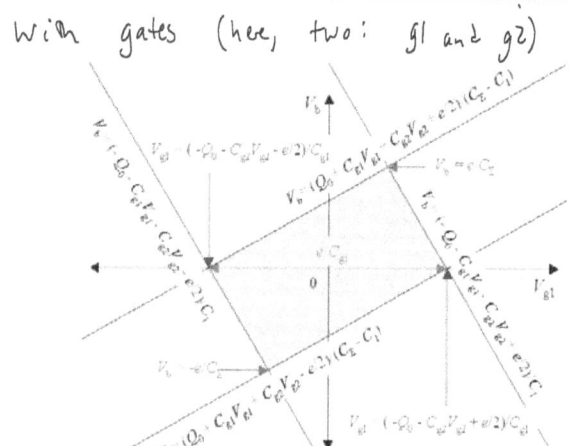
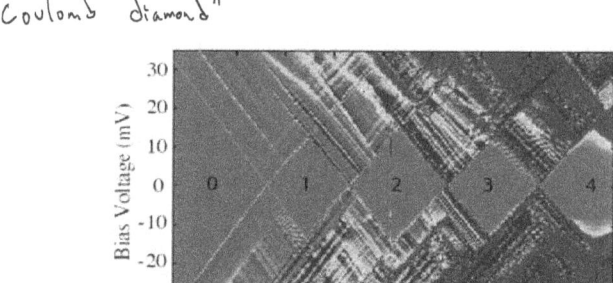

Quantum Hall Effect (Integer/ 2DEG)

Magnetic field enters the Hamiltonian thru canonical momentum: $\frac{\vec{p}^2}{2m} \to \frac{(\vec{p}-e\vec{A})^2}{2m}$

For $\vec{B} = B\hat{z}$, we choose symmetric gauge $\vec{A} = \frac{B}{2}(x\hat{y} - y\hat{x})$ (no spin) (no scattering) $\omega_c \tau \gg 1$

$$\vec{B} = \vec{\nabla} \times \vec{A} = \begin{vmatrix} \hat{x} & \hat{y} & \hat{z} \\ \partial_x & \partial_y & \partial_z \\ -\frac{By}{2} & \frac{Bx}{2} & 0 \end{vmatrix} = B\hat{z} \checkmark$$

Then $H = \frac{(p_x + \frac{eBy}{2})^2 + (p_y - \frac{eBx}{2})^2}{2m} = \frac{p_x^2 + p_y^2}{2m} + \frac{1}{2m}\frac{e^2B^2}{4m^2}(x^2+y^2) - \frac{eB}{2m}(p_y x - p_x y)$

$\qquad = \frac{\vec{p}^2}{2m} + \frac{1}{2}m\left(\frac{\omega_c}{2}\right)^2(x^2+y^2) - \frac{\omega_c}{2}L_z$

Immediate conclusions based on knowledge of harmonic oscillator:

→ Ladder of equally-spaced states, $\Delta E = 2 \times \frac{\hbar \omega_c}{2} = \hbar \omega_c$ "Landau levels"

→ $\psi \propto \exp\left(-\frac{(x^2+y^2)}{2 l_0^2}\right)$ where $l_0 = \sqrt{\frac{2\hbar}{m\omega_c}}$ sets lengthscale

Filling Fraction

How many "Landau Levels" are filled?

All states within $\Delta E = \hbar \omega_c$ condense into one LL.

In 2D, density of states

$$\frac{d^2k}{(2\pi)^2} \to \frac{2\pi k\, dk}{(2\pi)^2} = \frac{1}{2\pi}\sqrt{\frac{2mE}{\hbar^2}}\frac{\sqrt{\frac{2m}{\hbar^2 E}}}{2}dE = \frac{m}{2\pi\hbar^2}dE$$

So DOS is constant!

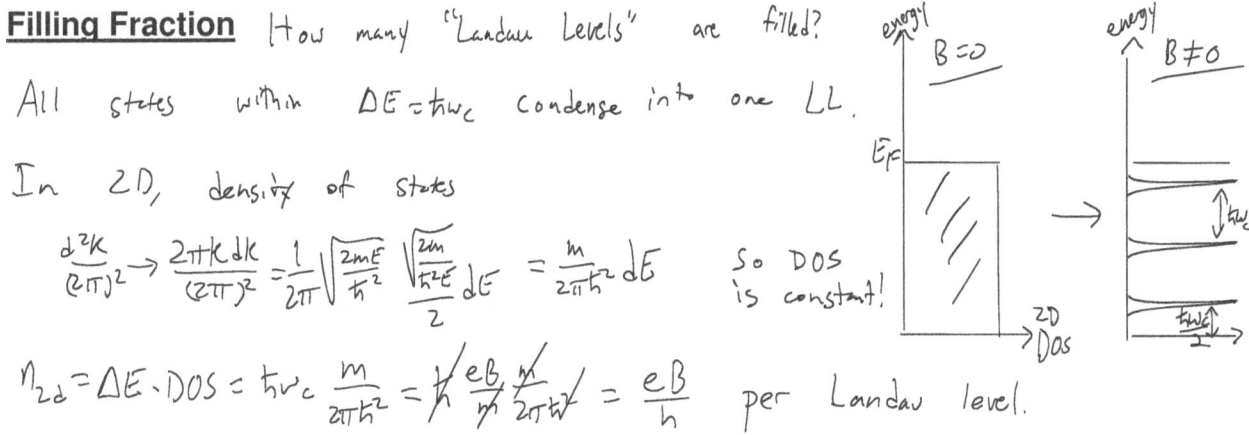

$$n_{2d} = \Delta E \cdot DOS = \hbar \omega_c \frac{m}{2\pi\hbar^2} = \hbar \frac{eB}{m}\frac{m}{2\pi\hbar^2} = \frac{eB}{h} \text{ per Landau level.}$$

So Hall resistance

$$R_{Hall} = \frac{V_{Hall}}{I} = -\frac{B}{e(n\,t)} = -\frac{B}{e\left(\frac{eB}{h}\nu\right)} = -\frac{h}{e^2}\frac{1}{\nu}$$

n_{2d} ↑

"filling fraction" ($\#$ of filled Landau levels)

quantum of resistance $\sim 26\,k\Omega$ "von Klitzing"

Experiment (1980)

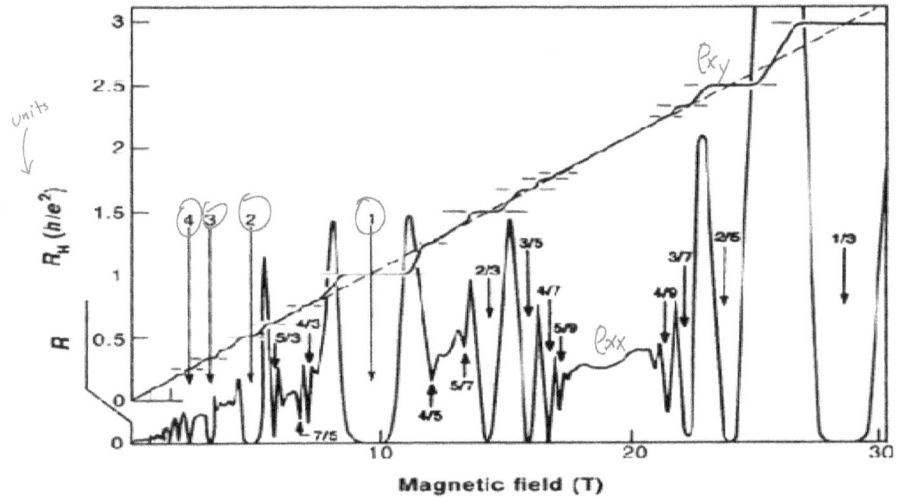

Note constant Hall resistivity ρ_{xy} in units of $\frac{h}{e^2\nu}$ $\nu = 1, 2, 3 \ldots$

Von Klitzing won 1985 Nobel prize for discovery of <u>exact</u> Hall resistance quantization

Edge states and ρ_{xx}

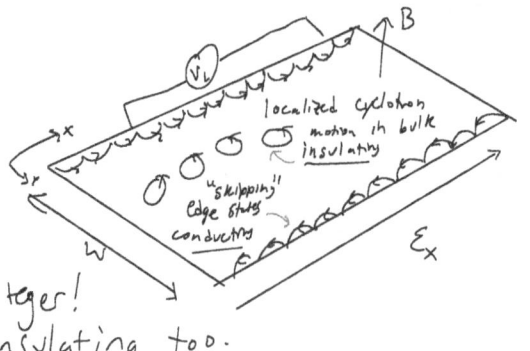

How can longitudinal resistivity $\rho_{xx} \to 0$ within a quantum hall plateau?

Because backscattering a chiral state requires transport across sample to opposite edge $\longrightarrow$ and there are no available states at E_F when ν is an integer! For the same reason, the bulk is insulating too.

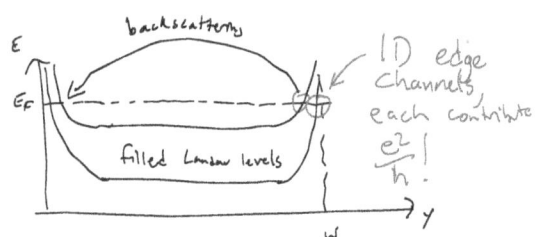

Since σ, ρ are tensors, $\rho_{xx}=0$ and σ_{xx} simultaneously:

$$\hat{\sigma} = \hat{\rho}^{-1} = \begin{bmatrix} 0 & \rho_{xy} \\ -\rho_{xy} & 0 \end{bmatrix}^{-1} = \begin{bmatrix} 0 & -\frac{1}{\rho_{xy}} \\ \frac{1}{\rho_{xy}} & 0 \end{bmatrix}$$

(arrow to upper-left 0: σ_{xx})

Cyclotron orbits have area $A = \pi l_0^2 = \pi \left(\sqrt{\frac{2\hbar}{m\omega_c}}\right)^2$

magnetic flux per orbit $\Phi = B \cdot A = \cancel{B} \frac{2\pi\hbar}{m \frac{eB}{m}} = \frac{h}{e}$ "flux quantum"

Could have predicted this on dimensionality alone!

Landau levels in 3D

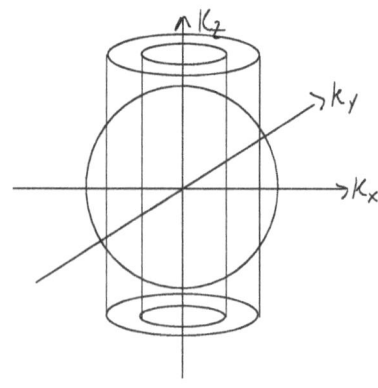

Applying $\vec{B} = B\hat{z}$ quantizes energy associated with motion in the x-y plane (cyclotron orbits). States inside Fermi sphere must collapse onto "Landau tubes" where $k_x^2 + k_y^2$ is constant.

As B increases, these tubes leave the Fermi sphere and the occupation at the Fermi surface changes. This causes essentially all the physical/thermodynamic properties to oscillate w/ $\frac{1}{B}$.

How many Landau tubes intersect Fermi sphere of typical 3D metal?

$$\nu = \frac{E_F}{\hbar \omega_c} = \frac{E_F}{\hbar \frac{eB}{m}} = \frac{E_F}{2\mu_B B} = \frac{10\,eV}{2 \cdot 5.8 \times 10^{-5} eV/T \cdot B} \to \sim 10^5 \text{ in } 1T$$

This is a huge number, far exceeding $\nu \sim 1$ in 2DEG Quantum Hall regime. But we can still observe unit changes in high B fields!

Shubnikov-deHaas/deHaas-vanAlphen effect (oscillations in σ or χ)

As we change B, filling fraction changes in proportion to the cross-sectional area of the Fermi surface perpendicular to B:

$$\nu = \frac{E_F}{\hbar \omega_c} = \frac{\hbar^2 k_F^2 / 2m}{\hbar \frac{e}{m} B} = \frac{\hbar}{2\pi e}\left(\pi k_F^2\right)\frac{1}{B} = \left(\frac{S_F}{2\pi}\frac{\hbar}{e}\right)\frac{1}{B}$$

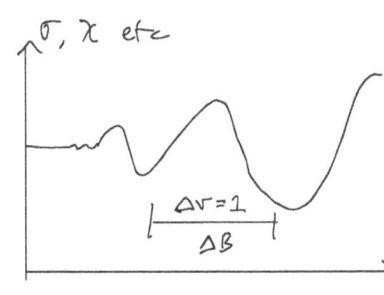

magnitude of oscillation becomes larger as filling fraction decreases and losing each LL makes a larger relative change.

We can find the period of oscillation in B:

$$\frac{1}{B} = \frac{2\pi e}{S_F \hbar}\nu \to \Delta\frac{1}{B} = \frac{2\pi e}{S_F \hbar}\Delta\nu = -\frac{\Delta B}{B^2} = \frac{2\pi e}{S_F \hbar} \to |\Delta B| = \frac{2\pi e}{S_F \hbar}B^2$$

This relationship is true even when the Fermi surface is not a sphere! By changing orientation of $\vec{B}$, anisotropy of S_F can be measured and Fermi surface reconstructed!

Thermodynamic properties in 3d : 3D-DOS

integrals over 3D k-space → integral over (1D) E

Cartesian infinitesimal $2\frac{d^3k}{(2\pi)^3} \xrightarrow{\text{spherical coords}} 2\frac{4\pi k^2 dk}{(2\pi)^3} \xrightarrow{E = \frac{\hbar^2 k^2}{2m}} \frac{1}{\pi^2}\left(\sqrt{\frac{2mE}{\hbar^2}}\right)^2\sqrt{\frac{2m}{\hbar^2 E}}\frac{dE}{2} = D(E)dE$

3D-Density of states $D(E) = \frac{1}{2\pi^2}\left(\frac{2m}{\hbar^2}\right)^{3/2} E^{1/2}$

A more convenient form:

Since $E_F = \frac{\hbar^2 k_F^2}{2m}$, $\frac{2m}{\hbar^2} = \frac{k_F^2}{E_F}$

$D(E) = \frac{1}{2\pi^2}\frac{1}{E_F}\left(\frac{E}{E_F}\right)^{1/2}(k_F^2)^{3/2} = \frac{1}{2\pi^2}\frac{1}{E_F}\left(\frac{E}{E_F}\right)^{1/2}\overbrace{(3\pi^2 n)}^{k_F^3} = \frac{3}{2}\frac{n}{E_F}\left(\frac{E}{E_F}\right)^{1/2}$

$\to$ for $E = E_F$, $D(E_F) = \frac{3}{2}\frac{n}{E_F}$

This is very useful since states near E_F dominate the thermodynamic properties of our electron system!

Application of 3D DOS @ E_F: magnetic susceptibility $\chi = \frac{M}{H}$ @ $T=0$

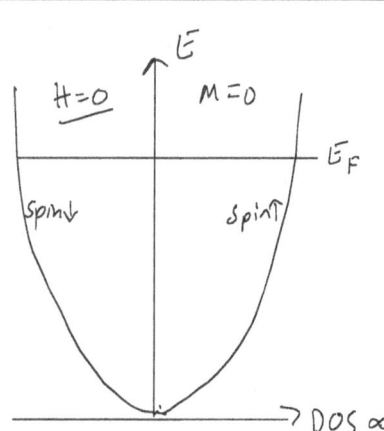

$$E_{zeeman} = -\vec{\mu} \cdot \vec{H}$$
$$= g \frac{\mu_B \vec{S}}{\hbar} \cdot \vec{H}$$
$$= \pm \mu_B H$$

(for spin ½ electrons, $g=2$ and $\langle\mu\rangle = \mu_B$)

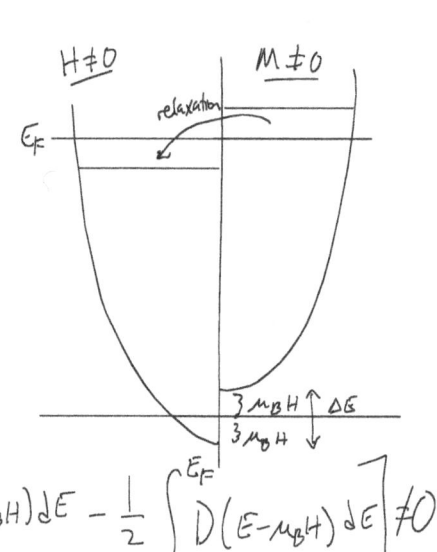

Magnetization $M = (n_\uparrow - n_\downarrow)\mu_B = \mu_B \left[\frac{1}{2}\int_{-\mu_B H}^{E_F} D(E+\mu_B H)dE - \frac{1}{2}\int_{+\mu_B H}^{E_F} D(E-\mu_B H)dE \right] \neq 0$

In linear response regime $H \to 0$ ($\Delta E \to 0$)

$M \sim \Delta E \frac{D(E_F)}{2} \mu_B = 2\mu_B H \cdot \frac{3}{2}\frac{n}{E_F} \mu_B = \left(\frac{3}{2} n \frac{\mu_B^2}{E_F}\right) H$ so $\chi = \frac{3}{2} n \frac{\mu_B^2}{E_F}$ "Pauli susceptibility"

Approximation of specific heat, $T \neq 0$

Fermi–Dirac occupation function $f(E) = \dfrac{1}{1 + e^{(E-\mu)/k_B T}}$ ← chemical potential

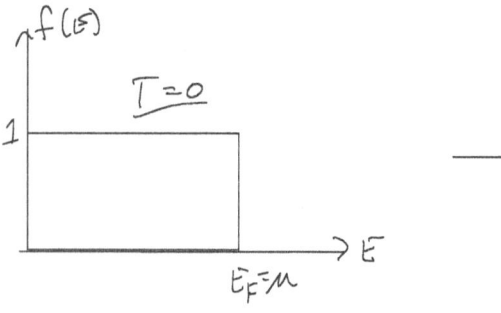
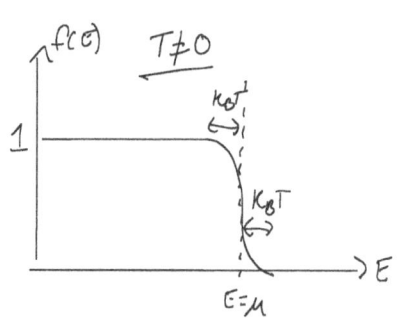

Change in energy density $\Delta U \sim \underbrace{k_B T \cdot D(E_F)}_{\text{\# of electrons affected}} \cdot \underbrace{k_B T}_{\text{energy gained}} = \frac{3}{2}\frac{n}{E_F}(k_B T)^2$

Specific heat $C_V = \left.\dfrac{dU}{dT}\right|_V = 3nk_B \dfrac{k_B T}{E_F} = 3nk_B \left(\dfrac{T}{T_F}\right)$ ← small @ RT

This explains why insulators have approx. same specific heat as metals: <u>not</u> due to electrons! But what? Stay tuned...

"Exact" Thermodynamic calculation for 3DEG

$\underset{T=0}{\underline{\quad\quad}} \quad U = \int_0^{E_F} D(E) E \, dE \quad \text{(easy!)} \quad \longrightarrow \quad \underset{T \neq 0}{\underline{\quad\quad}} \quad U = \int_0^\infty f(E) D(E) E \, dE = \int_0^\infty \frac{E D(E) dE}{1 + e^{(E-\mu)/k_B T}} \quad \text{(hard!)}$

"Sommerfeld expansion":

For $\int_{-\infty}^\infty G(E) f(E) \, dE$, define $K(E) = \int_{-\infty}^E G(E') dE'$ so that $G(E) = \frac{dK(E)}{dE}$

Then $\int_{-\infty}^\infty G(E) f(E) \, dE = \int_{-\infty}^\infty f \, dK \xrightarrow{\text{integration by parts}} \underbrace{fK\Big|_{-\infty}^\infty}_{\substack{f=0 \\ K=0}} - \int_{-\infty}^\infty K \frac{df}{dE} dE$

Since $\frac{df}{dE}$ is non-negligible only near $E=\mu$, Taylor expand $K(E)$:

$K(E) = K(\mu) + (E-\mu) \frac{dK}{dE}\Big|_{E=\mu} + \frac{(E-\mu)^2}{2!} \frac{d^2 K}{dE^2}\Big|_{E=\mu} + \cdots$

← all odd powers vanish by symmetry

$= K(\mu) \underbrace{\int_{-\infty}^\infty \left(-\frac{df}{dE}\right) dE}_{1} + \frac{dK}{dE}\Big|_{E=\mu} \underbrace{\int_{-\infty}^\infty \overset{\text{odd}}{(E-\mu)} \overset{\text{even}}{\left(-\frac{df}{dE}\right)} dE}_{\text{(about } E=\mu \text{)}}^{0} + \sum_{n=1}^\infty \frac{d^{2n} K}{dE^{2n}}\Big|_{E=\mu} \int_{-\infty}^\infty \frac{(E-\mu)^{2n}}{(2n)!} \left(-\frac{df}{dE}\right) dE$

To second order

$\int_{-\infty}^\infty G(E) f(E) dE = \overbrace{\int_{-\infty}^\mu G(E) dE}^{K(\mu)} + \overbrace{\frac{dG}{dE}\Big|_{E=\mu}}^{\frac{d^2 K}{dE^2}} \int_{-\infty}^\infty \frac{(E-\mu)^2}{2!} \left(-\frac{df}{dE}\right) dE + \cdots$

define unitless $x = \frac{E-\mu}{k_B T}$ Then $E-\mu = k_B T x$

$= \int_{-\infty}^\mu G(E) dE + (k_B T)^2 \frac{dG}{dE}\Big|_{E=\mu} \underbrace{\int_{-\infty}^\infty \frac{x^2}{2} \left(-\frac{d}{dx} \frac{1}{1+e^x}\right) dx}_{a_1 = \frac{\pi^2}{6}}$

So $\int_{-\infty}^\infty G(E) f(E) dE = \int_{-\infty}^\mu G(E) dE + \frac{\pi^2}{6} (k_B T)^2 \frac{dG}{dE}\Big|_{E=\mu} + \text{higher orders}$

Example at finite Temperature

electron density $\quad n = \int_{-\infty}^{\infty} D(E) f(E) dE$

since n conserved and $f(E) = \frac{1}{1+e^{(E-\mu)/k_B T}}$, this sets μ.

$G(E) = D(E)$:

$$n(T) = \int_{-\infty}^{\mu} D(E) dE + \frac{\pi^2}{6}(k_B T)^2 \frac{dD}{dE}\bigg|_{E=\mu} + \cdots$$

$$= \underbrace{\left[\int_0^{E_F} D(E) dE\right]}_{=n} + \underbrace{\left[\int_{E_F}^{\mu} D(E) dE + \frac{\pi^2}{6}(k_B T)^2 \frac{dD}{dE}\bigg|_{E=\mu} + \cdots\right]}_{=0!}$$

$$(\mu - E_F) D(E_F) + \frac{\pi^2}{6}(k_B T)^2 \frac{dD}{dE}\bigg|_{E=\mu} \approx 0$$

$$\mu \approx E_F - \frac{\pi^2}{6}(k_B T)^2 \frac{\frac{dD}{dE}}{D(E)}\bigg|_{E=\mu \approx E_F} \xrightarrow{D(E) \propto E^{1/2}} \mu \sim E_F - \frac{\pi^2}{6}(k_B T)^2 \left(\frac{1}{2 E_F}\right)$$
$$= E_F \left(1 - \frac{1}{3}\left(\frac{\pi k_B T}{2 E_F}\right)^2\right)$$

@ $T=0$, $\mu = E_F$ as expected and $\mu \approx E_F$ is a good approx!

Energy density and heat capacity

$$U = \int_{-\infty}^{\infty} E\, D(E) f(E) dE \qquad (G(E) = E \cdot D(E))$$

$$= \int_{-\infty}^{\mu} E\, D(E) dE + \frac{\pi^2}{6}(k_B T)^2 \frac{d}{dE}(E \cdot D(E))\bigg|_{E=\mu} + \cdots$$

$$= \int_0^{E_F} E\, D(E) dE + \int_{E_F}^{\mu} E\, D(E) dE + \frac{\pi^2}{6}(k_B T)^2 \left(D(E) + E \frac{dD}{dE}\right)\bigg|_{E=\mu} + \cdots$$

$$\cong U(T=0) + E_F \underbrace{\left[(\mu - E_F) D(E_F) + \frac{\pi^2}{6}(k_B T)^2 \frac{dD}{dE}\bigg|_{E=\mu}\right]}_{=0!\ (\text{see previous page})} + \frac{\pi^2}{6}(k_B T)^2 D(E_F)$$

So
$$U(T) \sim U_0 + \frac{\pi^2}{6}(k_B T)^2 D(E_F)$$

and
$$C_V = \frac{dU}{dT}\bigg|_V = \frac{\pi^2}{3} k_B^2 T D(E_F) = \frac{\pi^2}{2} \frac{k_B T}{E_F} n k_B = \frac{\pi^2}{2} n k_B \left(\frac{T}{T_F}\right)$$

cf. 3 in approximate method

Thermal properties redux

Wiedeman-Franz Law

$$\frac{K}{\sigma} = \frac{\frac{1}{3}v^2 \tau C_v}{\frac{ne^2\tau}{m}} = \frac{\frac{1}{3}\left(\frac{2E_F}{m}\right)\frac{\pi^2}{2}\frac{k_B T}{E_F}nk_B}{\frac{ne^2}{m}} = \frac{\pi^2}{3}\left(\frac{k_B}{e}\right)^2 T \quad \text{c.f. Drude } \frac{3}{2}\left(\frac{k_B}{e}\right)^2 T$$

$$= LT \sim \left(2.4 \times 10^{-8} \frac{W\Omega}{K^2}\right) T$$

Thermopower

We previously found $\mathcal{E} = -\frac{1}{3e}\frac{dE}{dT}\nabla T = -\frac{1}{3en}\frac{d(nE)}{dT}\nabla T = -\frac{C_v}{3en}\nabla T = \mathbb{K}T$

So $\mathbb{K} = -\frac{C_v}{3en} = -\frac{\frac{\pi^2}{2}\frac{k_B T}{E_F}nk_B}{3en} = -\frac{\pi^2}{6}\frac{k_B}{e}\left(\frac{T}{T_F}\right)$ c.f. Drude $-\frac{k_B}{2e}$

This smaller value is consistent w/ experiment and resolves the discrepancy w/ the Drude result. But where does the missing heat capacity come from?

Atom dynamics: model classical 1D "crystal"

diatomic linear chain

[diagram of diatomic linear chain with masses M_1, M_2, springs, showing $(n-1)^{th}$, n^{th}, $(n+1)^{th}$ unit cells, lattice constant a, and absolute coordinate x]

Classical kinetics given by Newton's 2nd Law: $F = ma$, $F = -f\Delta u$ ← spring constant

In n^{th} unit cell:

[diagram showing $u_{n-1,2}$, $u_{n,1}$, $u_{n,2}$, $u_{n+1,1}$]

$u_{n,\alpha}$ are relative coordinates → deviation from equilibrium

atom 1 (●): $-f(u_{n,1} - u_{n-1,2}) - f(u_{n,1} - u_{n,2}) = M_1 \ddot{u}_{n,1}$

atom 2 (○): $-f(u_{n,2} - u_{n+1,1}) - f(u_{n,2} - u_{n,1}) = M_2 \ddot{u}_{n,2}$

} coupled 2nd order diff. eqs.

Hermitian eigenvalue equation

Ansatz: plane wave solution $u_{n,\sigma} = \frac{1}{\sqrt{M_\sigma}} u_\sigma(q) e^{i(qx_n - \omega t)}$ $(x_n = na)$

upon substitution:

$-\frac{2f}{\sqrt{M_1}} u_1(q) e^{i(qna - \omega t)} + \frac{f}{\sqrt{M_2}} u_2(q) \left(e^{i(q(n-1)a - \omega t)} + e^{i(qna - \omega t)} \right) = -\sqrt{M_1} \omega^2 u_1(q) e^{i(qna - \omega t)}$

$-\frac{2f}{\sqrt{M_2}} u_2(q) e^{i(qna - \omega t)} + \frac{f}{\sqrt{M_1}} u_1(q) \left(e^{i(q(n+1)a - \omega t)} + e^{i(qna - \omega t)} \right) = -\sqrt{M_2} \omega^2 u_2(q) e^{i(qna - \omega t)}$

This is equivalent to the linear system:

$$\begin{bmatrix} \frac{2f}{M_1} & -\frac{f}{\sqrt{M_1 M_2}}(1 + e^{-iqa}) \\ -\frac{f}{\sqrt{M_1 M_2}}(1 + e^{+iqa}) & \frac{2f}{M_2} \end{bmatrix} \begin{bmatrix} u_1(q) \\ u_2(q) \end{bmatrix} = \omega^2 \begin{bmatrix} u_1(q) \\ u_2(q) \end{bmatrix}$$

note Hermitian symmetry so real eigenvalues (ω^2) are guaranteed!

Quadratic formula

$$\det \begin{vmatrix} \frac{2f}{M_1} - \omega^2 & -\frac{f}{\sqrt{M_1 M_2}}(1 + e^{-iqa}) \\ -\frac{f}{\sqrt{M_1 M_2}}(1 + e^{+iqa}) & \frac{2f}{M_2} - \omega^2 \end{vmatrix} = 0$$

Characteristic eqn. for eigenvalue ω^2:

$$(\omega^2)^2 - \left(\frac{2f}{M_1} + \frac{2f}{M_2}\right)\omega^2 + \frac{4f^2}{M_1 M_2} - \frac{f^2}{M_1 M_2} \underbrace{(2 + 2\cos qa)}_{4\cos^2 \frac{qa}{2}} = 0$$

Solution by quadratic formula: $\omega^2 = \dfrac{\frac{2f}{M_1} + \frac{2f}{M_2} \pm \sqrt{4f^2 \left(\frac{1}{M_1} + \frac{1}{M_2}\right)^2 - 4\left(\frac{4f^2}{M_1 M_2}(1 - \cos^2 \frac{qa}{2})\right)}}{2}$

$\omega^2 = f\left(\frac{1}{M_1} + \frac{1}{M_2}\right) \pm f\left[\left(\frac{1}{M_1} + \frac{1}{M_2}\right)^2 - \frac{4}{M_1 M_2} \sin^2 \frac{qa}{2}\right]^{1/2}$ (note $\frac{2\pi}{a}$ periodicity in q!)

Limits of dispersion

For $qa \ll 1$: approx. $\sin^2 \frac{qa}{2} \sim \frac{(qa)^2}{4}$ and expand sqrt to first order:
(long wavelength)

$$\omega^2 \sim f\left(\frac{1}{m_1}+\frac{1}{m_2}\right)\left(1 \pm \left(1 - \frac{1}{2} \frac{\cancel{4}}{m_1 m_2} \frac{(m_1 m_2)^2}{(m_1+m_2)^2} \frac{(qa)^2}{\cancel{4}}\right)\right) \sim \begin{cases} 2f\left(\frac{1}{m_1}+\frac{1}{m_2}\right) & (+) \\ \dfrac{f}{m_1+m_2} \dfrac{(qa)^2}{2} & (-) \end{cases}$$

For $q = \frac{\pi}{a}$: $\omega^2 = f\left(\frac{1}{m_1}+\frac{1}{m_2}\right) \pm f\left[\frac{(m_1+m_2)^2 - 4m_1 m_2}{(m_1 m_2)^2}\right]^{1/2} = f\left(\frac{1}{m_1}+\frac{1}{m_2}\right) \pm f\left(\frac{1}{m_1}-\frac{1}{m_2}\right)$

Full dispersion and eigenmodes

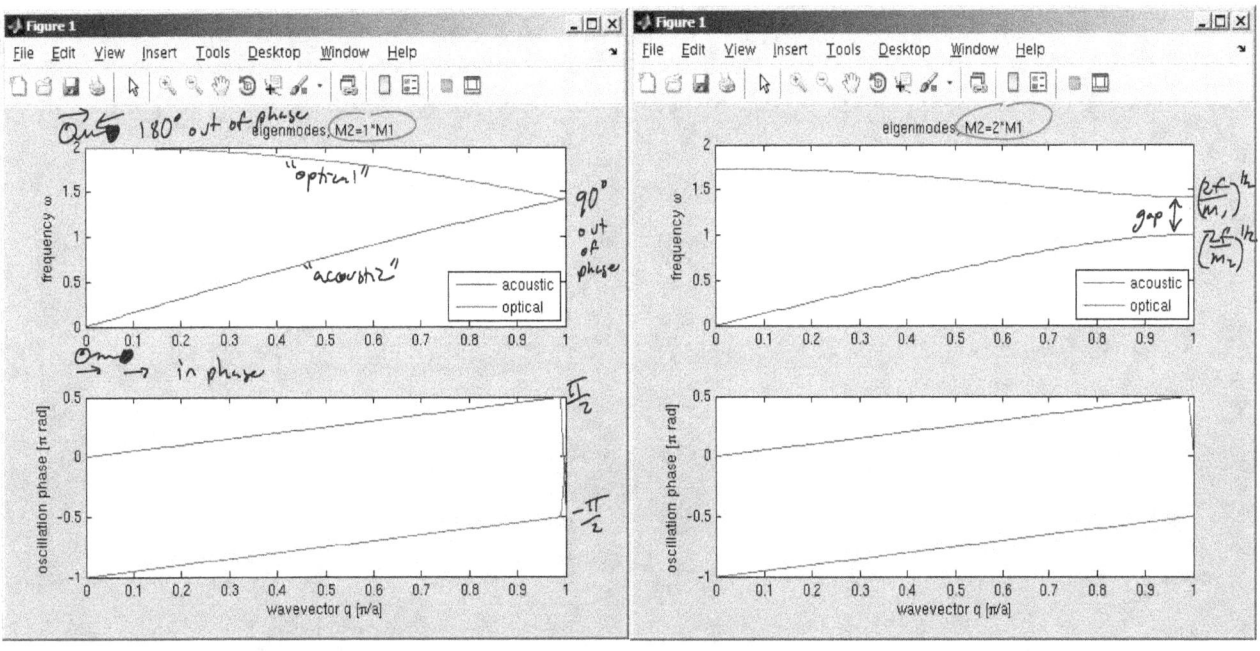

Extension to 3D

In 3D crystals, we still have "acoustic" and "optical" branches, but now displacement vector $\vec{u}_\alpha$ can be both longitudinal ($\vec{u} \parallel \vec{q}$) or transverse ($\vec{u} \perp \vec{q}$) to propagation direction. Therefore, we include 1 longitudinal and 2 transverse polarizations:

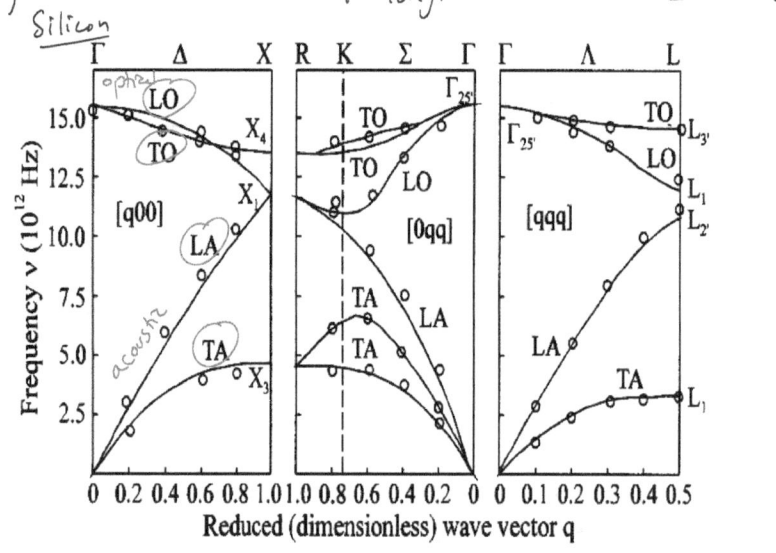

This dispersion is plotted along the perimeter of periodically repeated 3D "Brillouin zone" in $\vec{q}$-space connecting points of high symmetry → more on this later in the course!

Thermodynamics of coupled harmonic oscillators: QM

QM spectrum of single-particle harmonic oscillator: $E_n = \hbar\omega(n+\tfrac{1}{2})$ [$n = 0,1,2...$]

Occupation of each state n given by Boltzmann statistics $P_n \propto e^{-\frac{E_n}{k_BT}}$

Normalization: Total occupation probability is unity: $\sum_{n=0}^{\infty} P_n = 1$

$$\sum_n C e^{-\frac{\hbar\omega(n+\tfrac{1}{2})}{k_BT}} = C e^{-\frac{\hbar\omega}{2k_BT}} \sum_{n=0}^{\infty} e^{-\frac{\hbar\omega n}{k_BT}} \longrightarrow \text{a geometric series!}$$

$$\left[1 + x + x^2 + \cdots = 1 + x(1 + x + x^2 + \cdots) \longrightarrow \sum_n x^n = 1 + x\sum_n x^n \text{ so } \sum_n x^n = \frac{1}{1-x}\right]$$

Then $C e^{-\frac{\hbar\omega}{2k_BT}} \frac{1}{1 - e^{-\hbar\omega/k_BT}} = 1$ so $C = \frac{1 - e^{-\hbar\omega/k_BT}}{e^{-\hbar\omega/2k_BT}} = e^{\frac{\hbar\omega}{2k_BT}}(1 - e^{-\hbar\omega/k_BT})$

Normalized occupation probability $P_n = e^{\frac{\hbar\omega}{2k_BT}}(1 - e^{-\hbar\omega/k_BT}) e^{-\frac{\hbar\omega}{k_BT}(n+\tfrac{1}{2})}$

$= e^{-\frac{\hbar\omega n}{k_BT}}(1 - e^{-\hbar\omega/k_BT})$

Energy of harmonic oscillator ensemble

Each mode of the linear chain is an oscillator w/ frequency ω. How much thermal energy does this mode have?

Expectation value of energy

$$\langle E \rangle = \sum_{n=0}^{\infty} P_n E_n = \sum_n \left[e^{-\frac{\hbar\omega n}{k_B T}} \left(1 - e^{-\frac{\hbar\omega}{k_B T}}\right) \right] \hbar\omega\left(n + \tfrac{1}{2}\right) = \hbar\omega\left(1 - e^{-\frac{\hbar\omega}{k_B T}}\right) \sum_n \left(n + \tfrac{1}{2}\right) e^{-\frac{\hbar\omega n}{k_B T}}$$

Since $\sum_n x^n = \frac{1}{1-x}$, $\frac{d}{dx}\sum_n x^n = \sum_n n x^{n-1} = \frac{1}{(1-x)^2}$ so $\sum_n n x^n = \frac{x}{(1-x)^2}$

$$\langle E \rangle = \hbar\omega\left(1 - e^{-\frac{\hbar\omega}{k_B T}}\right) \left[\frac{e^{-\hbar\omega/k_B T}}{(1-e^{-\hbar\omega/k_B T})^2} + \frac{1}{2} \frac{1}{1-e^{-\hbar\omega/k_B T}} \right] = \hbar\omega \left(\frac{e^{-\hbar\omega/k_B T}}{1-e^{-\hbar\omega/k_B T}} + \tfrac{1}{2} \right)$$

$$= \hbar\omega\left(\frac{1}{e^{\hbar\omega/k_B T} - 1} + \tfrac{1}{2} \right) = \hbar\omega\left(\langle n \rangle + \tfrac{1}{2} \right)$$

↗ Bose-Einstein occupation

↑ number of "phonons": fictitious, non-interacting, non-conserved "particles"

So we can think of this problem in terms of a phonon gas!

Phonon energy density

$$U = \iiint \langle E \rangle \frac{d^3 q}{(2\pi)^3} \qquad \text{But } \langle E \rangle \text{ is a function of } \omega \text{ not } q!$$
$$|\vec{q}| < q_{max} \approx \tfrac{\pi}{a}$$

# of atoms per unit cell ↓
# of unit cells ↓

$$N = V \iiint_{|\vec{q}|<q_{max}} \frac{d^3 q}{(2\pi)^3} \longrightarrow V \int_0^{q_{max}} \frac{4\pi q^2 dq}{(2\pi)^3} \xrightarrow{q = \omega/c \text{ acoustic dispersion}} \frac{V}{2\pi^2} \int_0^{\omega_{max} = \omega_D} \frac{\omega^2 d\omega}{c^3} = \frac{V}{2\pi^2} \frac{\omega_D^3}{3c^3}$$

"Debye frequency"

So each branch of acoustic phonons has a freq DOS $= \frac{1}{2\pi^2} \frac{\omega^2}{c^3}$ $\left[\frac{1}{\text{vol}} \text{ per Hz}\right]$

Energy density of 1 longitudinal and 2 transverse modes in 3D:

$$U = \frac{1}{2\pi^2}\left(\frac{1}{c_L^3} + \frac{2}{c_T^3}\right) \int_0^{\omega_D} \hbar\omega \left(\frac{1}{e^{\hbar\omega/k_B T}-1} + \tfrac{1}{2}\right) \omega^2 d\omega$$

Phonon specific heat capacity

$$C_V = \frac{dU}{dT} = \frac{\hbar}{2\pi^2}\left(\frac{1}{c_L^3} + \frac{2}{c_T^3}\right)\int_0^{\omega_D} \frac{e^{\hbar\omega/k_B T}}{(e^{\hbar\omega/k_B T}-1)^2} \frac{\hbar\omega}{k_B T^2} \omega^3 d\omega$$

From normalization, $\frac{3rN}{V} = \frac{1}{2\pi^2}\left(\frac{1}{c_L^3} + \frac{2}{c_T^3}\right)\frac{\omega_D^3}{3}$

So $\frac{1}{2\pi^2}\left(\frac{1}{c_L^3} + \frac{2}{c_T^3}\right) = \frac{3rN}{V}\frac{3}{\omega_D^3}$, giving

$$C_V = \hbar \frac{3rN}{V}\frac{3}{\omega_D^3} \int_0^{\omega_D} \frac{e^{\hbar\omega/k_B T}}{(e^{\hbar\omega/k_B T}-1)^2} \frac{\hbar\omega}{k_B T^2} \omega^3 d\omega$$

Now, substitute $y = \frac{\hbar\omega}{k_B T}$ so that $\omega = \frac{y k_B T}{\hbar}$, $d\omega = \frac{k_B T}{\hbar} dy$, $\omega_D = \frac{k_B T y_{max}}{\hbar}$

$$C_V = \hbar \frac{3rN}{V}\frac{3}{\omega_D^3} \int_0^{y_{max} = \frac{\hbar\omega_D}{k_B T} = \frac{\Theta}{T} \text{ "Debye temperature"}} \frac{e^y}{(e^y-1)^2} \frac{y}{T} y^3 \left(\frac{k_B T}{\hbar}\right)^4 dy = \frac{3rN k_B}{V} \cdot 3\left(\frac{T}{\Theta}\right)^3 \int_0^{\Theta/T} \frac{e^y y^4}{(e^y-1)^2} dy$$

High-temperature limit

$k_B T \gg \hbar\omega_D$ $(T \gg \Theta, y_{max} \to 0)$ So Taylor expand $e^y \approx 1+y$

To lowest order

$$C_V \approx \frac{3rN k_B}{V} \cdot 3\left(\frac{T}{\Theta}\right)^3 \int_0^{\Theta/T} y^2 dy = \frac{3rN k_B}{V} \cdot 3\left(\frac{T}{\Theta}\right)^3 \frac{1}{3}\left(\frac{\Theta}{T}\right)^3 = 3n k_B$$

Matches Dulong-Petit! However, it is clearly due to mechanical degrees of freedom of the **coupled** atomic lattice, rather than the independent motion of classical charged electrons as assumed by Drude!

Low-temperature limit

In low T limit, $T \ll \theta$, $y_{max} = \frac{\theta}{T} \to \infty$

$$\sim \int_0^\infty \frac{e^y y^4}{(e^y-1)^2} dy = \frac{4\pi^4}{15} \quad \text{no temp. dependence!}$$

$$C_V \cong \frac{3 \cdot N k_B}{V} \left(\frac{T}{\theta}\right)^3 \frac{4\pi^4}{15}$$

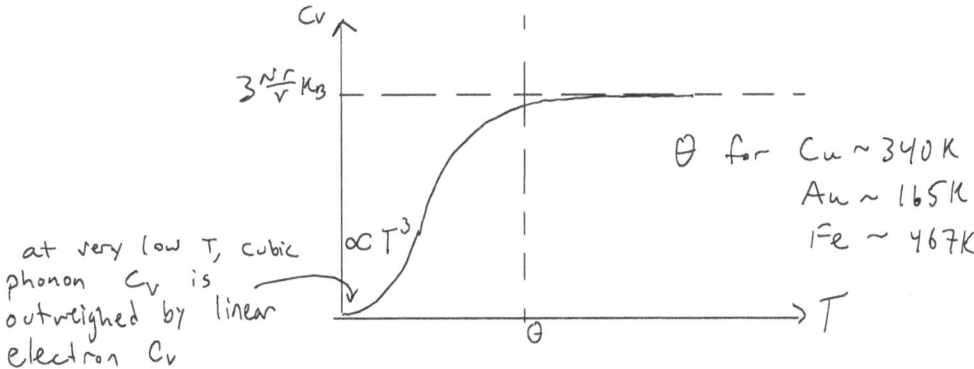

at very low T, cubic phonon C_V is outweighed by linear electron C_V

θ for Cu ~ 340 K
Au ~ 165 K
Fe ~ 467 K

Beyond the "noninteracting" approximation: screening

Consider a charged impurity in the otherwise homogeneous 3D electron gas:

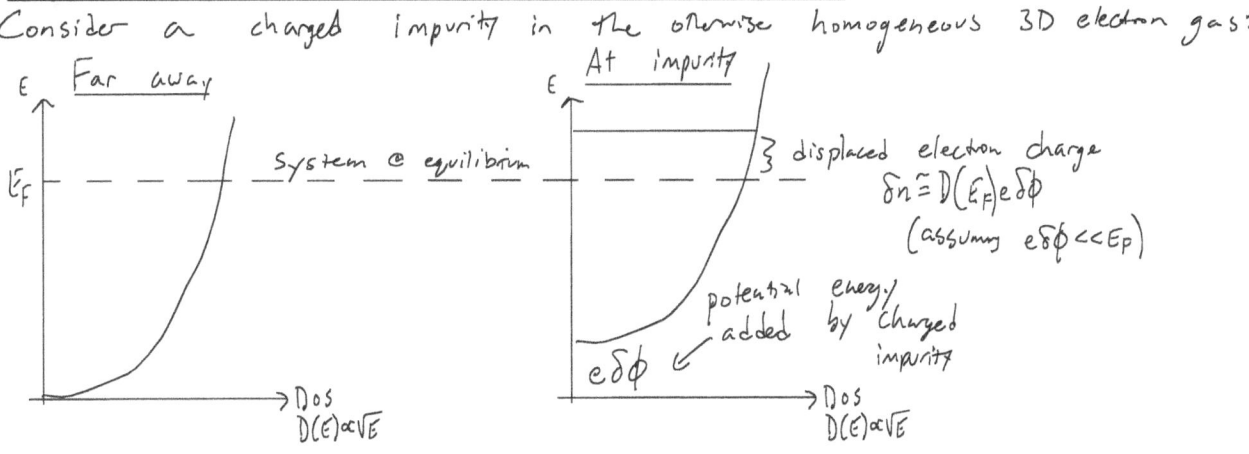

displaced electron charge
$\delta n \cong D(E_F) e \delta\phi$
(assuming $e\delta\phi \ll E_F$)

potential energy added by charged impurity

At intermediate locations, solve Poisson's equation:

$$\vec{\nabla}^2 \delta\phi(\vec{r}) = \frac{\rho}{\epsilon_0} = \frac{e \delta n}{\epsilon_0} = \frac{e^2 \delta\phi D(E_F)}{\epsilon_0} = \lambda^2 \delta\phi(\vec{r}) \quad \left(\lambda \equiv \sqrt{\frac{e^2 D(E_F)}{\epsilon_0}}, [\lambda] = \text{length}^{-1}\right)$$

Just as we solve $\vec{\nabla}^2 \phi = -\delta(\vec{r})$ to determine Green's function $\phi(\vec{r}) = \frac{1}{4\pi |r|}$, here we must solve

$$\vec{\nabla}^2 (\delta\phi(\vec{r})) - \lambda^2 (\delta\phi(\vec{r})) = -\delta(\vec{r})$$

Green's function solution

Use Fourier transform $\delta\phi(r) = \frac{1}{(2\pi)^3}\int \widetilde{\delta\phi}(k) e^{i\vec{k}\cdot\vec{r}} d^3k$ so that

$$\nabla^2 \delta\phi = \frac{1}{(2\pi)^3}\int (-k^2)\widetilde{\delta\phi} e^{i\vec{k}\cdot\vec{r}} d^3k \quad \text{and} \quad \delta(\vec{r}) = \frac{1}{(2\pi)^3}\int e^{i\vec{k}\cdot\vec{r}} d^3k$$

Substitute into Poisson's equation and project onto $e^{i\vec{k}'\cdot\vec{r}}$ orthonormal basis

$$-k^2 \widetilde{\delta\phi} - \lambda^2 \widetilde{\delta\phi} = -1 \quad \longrightarrow \quad \widetilde{\delta\phi} = \frac{1}{k^2 + \lambda^2}$$

So
$$\delta\phi(\vec{r}) = \frac{1}{(2\pi)^3}\int \frac{e^{i\vec{k}\cdot\vec{r}}}{k^2 + \lambda^2} d^3k$$
all k-space.

This integrand does not look friendly to cartesian coordinates! $\vec{k}\cdot\vec{r} = |k||r|\cos\theta$ suggests we use spherical coords instead.

Spherical coordinates

In spherical coordinates, $d^3k = k^2 \sin\theta\, d\theta\, d\varphi\, dk$, $\vec{k}\cdot\vec{r} = kr\cos\theta$
(align polar axis parallel to real-space coordinate $\vec{r}$)

$$\delta\phi(r) = \frac{2\pi}{(2\pi)^3}\int_0^\infty \int_0^\pi \frac{k^2 e^{ikr\cos\theta}}{k^2+\lambda^2}\sin\theta\, d\theta\, dk = \frac{1}{(2\pi)^2}\int_0^\infty \frac{k^2}{k^2+\lambda^2}\left(-\frac{e^{ikr\cos\theta}}{ikr}\bigg|_0^\pi\right)dk$$

$$= \frac{1}{(2\pi)^2}\int_0^\infty \frac{k^2}{k^2+\lambda^2}\left(\frac{e^{ikr} - e^{-ikr}}{ikr}\right)dk$$

Now, split the integral

$$= \frac{1}{i(2\pi)^2 r}\left[\int_0^\infty \frac{k e^{ikr}}{k^2+\lambda^2} dk - \int_0^\infty \frac{k e^{-ikr}}{k^2+\lambda^2} dk\right] \xrightarrow[\text{in 2nd integral}]{k \to -k} \frac{1}{i(2\pi)^2 r}\left[\int_0^\infty \frac{k e^{ikr}}{k^2+\lambda^2} dk - \int_0^{-\infty} \frac{k e^{ikr}}{k^2+\lambda^2} dk\right]$$

Same integrand!

invert bounds on second integral:

$$= \frac{1}{i(2\pi)^2 r}\int_{-\infty}^{+\infty} \frac{k e^{ikr}}{k^2+\lambda^2} dk \qquad \text{How to evaluate this?}$$

Note that integrand $\to 0$ as $|k| \to \infty$.

Contour Integration

$$\frac{1}{i(2\pi)^2 r}\int_{-\infty}^{\infty}\frac{K e^{iKr}dK}{(K+i\lambda)(K-i\lambda)}$$

Close contour in complex plane → enclose pole

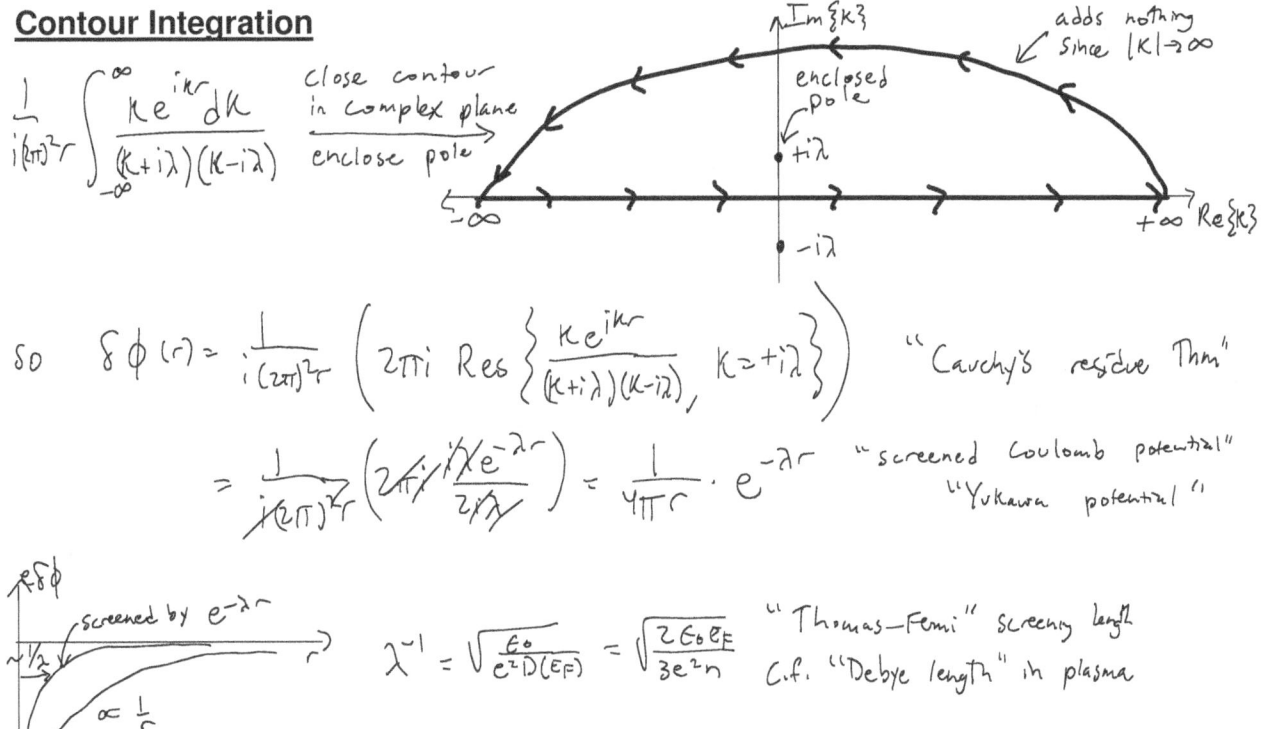

adds nothing since $|K|\to\infty$
enclosed pole $+i\lambda$

so $\delta\phi(r) = \dfrac{1}{i(2\pi)^2 r}\left(2\pi i \operatorname{Res}\left\{\dfrac{Ke^{iKr}}{(K+i\lambda)(K-i\lambda)}, K=+i\lambda\right\}\right)$ "Cauchy's residue Thm"

$= \dfrac{1}{i(2\pi)^2 r}\left(2\pi i \dfrac{i\lambda e^{-\lambda r}}{2i\lambda}\right) = \dfrac{1}{4\pi r}\cdot e^{-\lambda r}$ "screened Coulomb potential" / "Yukawa potential"

$\lambda^{-1} = \sqrt{\dfrac{\varepsilon_0}{e^2 D(E_F)}} = \sqrt{\dfrac{2\varepsilon_0 E_F}{3e^2 n}}$ "Thomas–Fermi" screening length

c.f. "Debye length" in plasma

screened by $e^{-\lambda r}$, $\propto \frac{1}{r}$

→ Screening attractive potentials is especially important when electron density is high enough to eliminate bound states → "metal–insulator transition"

Interactions in the electron gas: Ordered magnetic states

First, local moments, no interactions $\mathcal{H} = -\sum_i^N \vec{\mu}_i\cdot\vec{H}$ (spin ½)

$M = N\langle\mu\rangle = N\dfrac{\sum \mu_i e^{-E_i/k_BT}}{\sum e^{-E_i/k_BT}} = N\dfrac{+\mu_B e^{-(-\mu_B H)/k_BT} + (-\mu_B)e^{-\mu_B H/k_BT}}{e^{-(-\mu_B H)/k_BT} + e^{-\mu_B H/k_BT}} = N\mu_B \tanh\dfrac{\mu_B H}{k_BT}$

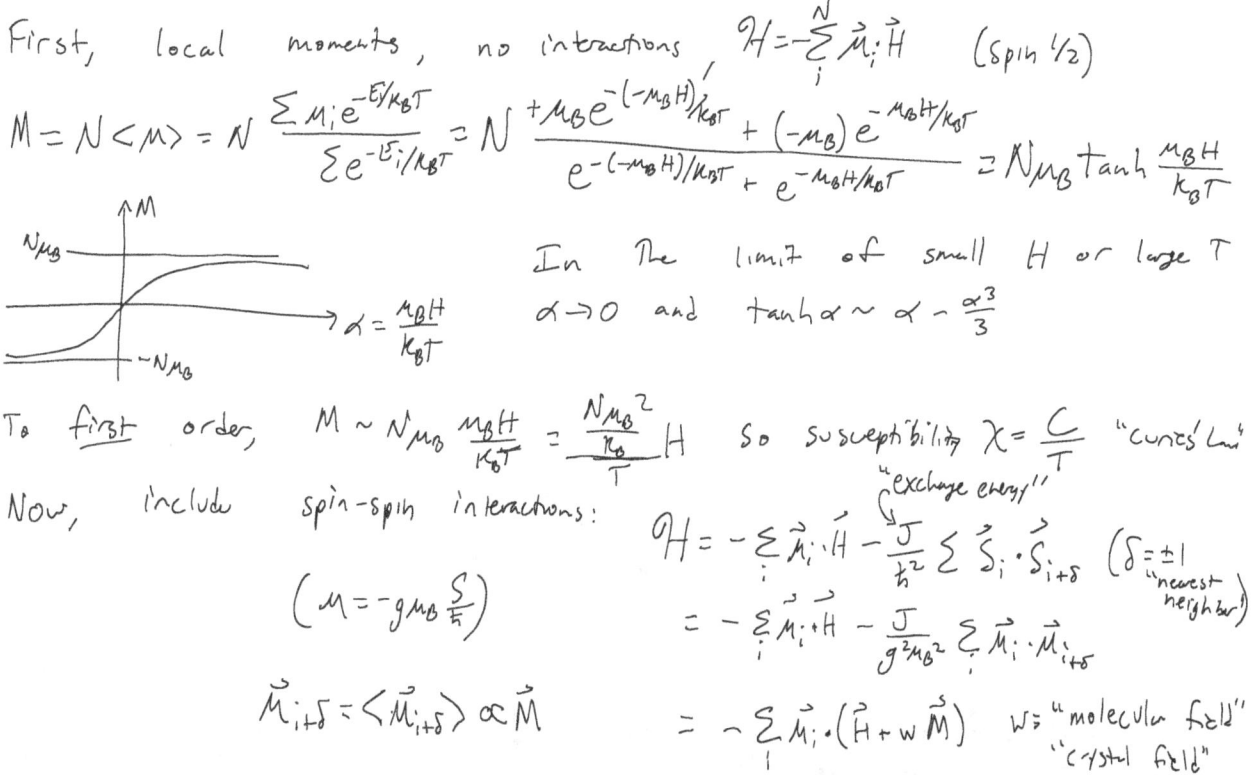

In the limit of small H or large T, $\alpha = \dfrac{\mu_B H}{k_BT}$, $\alpha \to 0$ and $\tanh\alpha \sim \alpha - \dfrac{\alpha^3}{3}$

To first order, $M \sim N\mu_B \dfrac{\mu_B H}{k_BT} = \dfrac{N\mu_B^2/k_B}{T}H$ So susceptibility $\chi = \dfrac{C}{T}$ "Curie's Law"

Now, include spin-spin interactions: $\mathcal{H} = -\sum_i \vec{\mu}_i\cdot\vec{H} - \dfrac{J}{\hbar^2}\sum \vec{S}_i\cdot\vec{S}_{i+\delta}$ ($\delta = \pm 1$ "nearest neighbor")

"exchange energy"

$\left(\mu = -g\mu_B \dfrac{S}{\hbar}\right)$

$= -\sum_i \vec{\mu}_i\cdot\vec{H} - \dfrac{J}{g^2\mu_B^2}\sum_i \vec{\mu}_i\cdot\vec{\mu}_{i+\delta}$

$\vec{\mu}_{i+\delta} = \langle\vec{\mu}_{i+\delta}\rangle \propto \vec{M}$

$= -\sum_i \vec{\mu}_i\cdot(\vec{H} + w\vec{M})$ w = "molecular field" / "crystal field"

Ferromagnetism ($M \neq 0$ even when $H=0$!)

$$M = N\mu_B \tanh \alpha' = N\mu_B \left(\alpha' - \frac{\alpha'^3}{3} + \dots\right), \quad \alpha' \equiv \frac{\mu_B}{k_B T}(H + wM)$$

If $H=0$, $\gamma \equiv \frac{\mu_B w}{k_B T}$, $\alpha' = \gamma M$

$$M \approx N\mu_B \left(\gamma M - \frac{\gamma^3 M^3}{3}\right) \longrightarrow M = \left(\frac{3}{\gamma^3}\left(\gamma - \frac{1}{N\mu_B}\right)\right)^{1/2}$$

for $\gamma > \frac{1}{N\mu_B}$, M is nonzero, real: $\frac{\mu_B w}{k_B T} > \frac{1}{N\mu_B} \rightarrow T < \frac{N\mu_B^2 w}{k_B} \equiv T_c$ "Curie temp"

Just below T_c, $M = \frac{\sqrt{3}}{\gamma}\left(1 - \frac{1}{N\gamma\mu_B}\right)^{1/2} = \frac{\sqrt{3}}{\gamma}\left(1 - \frac{k_B T}{N\mu_B^2 w}\right)^{1/2} = \frac{\sqrt{3}}{\gamma}\left(1 - \frac{T}{T_c}\right)^{1/2}$

T_c:
Ni : 358°C
Fe : 770°C
Co : 1130°C

Above T_C: paramagnetic phase

$$M = N_\mu \tanh \alpha' \approx N_\mu \frac{(H+wM)\mu_B}{k_B T} \quad \text{to first order}$$

Solution:

$$M = \frac{\frac{N\mu_B^2}{k_B}}{T - \frac{N\mu_B^2 w}{k_B}} H = \left(\frac{C}{T - T_c}\right) H \longrightarrow \chi = \frac{C}{T - T_c} \quad \text{"Curie-Weiss" Law}$$

But, what does this have to do w/ the free electron gas? How does exchange interaction of <u>delocalized</u> electron planewaves lead to ferromagnetism?

Exchange energy in the free electron gas

2-electron pair wavefunction must be totally **antisymmetric**, so for aligned spins, must be spatially **anti-symmetric**

$$\Psi_{ij} \propto \psi_{k_i}(r_i)\psi_{k_j}(r_j) - \psi_{k_i}(r_j)\psi_{k_j}(r_i)$$

$$\propto e^{i\vec{k}_i \cdot \vec{r}_i} e^{i\vec{k}_j \cdot \vec{r}_j} - e^{i\vec{k}_i \cdot \vec{r}_j} e^{i\vec{k}_j \cdot \vec{r}_i} = e^{i\vec{k}_i \cdot \vec{r}_i} e^{i\vec{k}_j \cdot \vec{r}_j}\left(1 - e^{-i(\vec{k}_i - \vec{k}_j)\cdot(\vec{r}_i - \vec{r}_j)}\right)$$

Pairwise probability density

$$|\Psi_{ij}|^2 \propto 1 - \cos\left((\vec{k}_i - \vec{k}_j)\cdot\vec{r}\right), \quad (\vec{r} = \vec{r}_i - \vec{r}_j)$$

$$\rho_\uparrow = \frac{ne}{2} \frac{\iint [1 - \cos((\vec{k}_i - \vec{k}_j)\cdot\vec{r})] d^3k_i \, d^3k_j}{(\frac{4}{3}\pi k_F^3)^2} \quad \left(\begin{array}{c}\cos(\alpha-\beta) = \overbrace{\cos\alpha\cos\beta}^{\text{even}} + \overbrace{\sin\alpha\sin\beta}^{\text{odd}} \\ \text{so symmetric integration leads to zero} \\ \text{contribution from second term.}\end{array}\right)$$

$$= \frac{ne}{2}\left[1 - \frac{\int \cos\vec{k}_i\cdot\vec{r}\, d^3k_i \int \cos\vec{k}_j\cdot\vec{r}\, d^3k_j}{(\frac{4}{3}\pi k_F^3)^2}\right] = \frac{ne}{2}\left[1 - \left(\frac{\int \cos\vec{k}\cdot\vec{r}\, d^3k}{\frac{4}{3}\pi k_F^3}\right)^2\right]$$

Fourier transform of the Fermi sphere

aligned polar axis to real-space coordinate

$$\int_{\text{Fermi sphere}} \cos\vec{k}\cdot\vec{r}\, d^3k = 2\pi \int_0^{k_F}\int_0^\pi \cos(kr\cos\theta)\sin\theta\, k^2\, d\theta\, dk$$

$$= 2\pi \int_0^{k_F}\left(-\frac{\sin(kr\cos\theta)}{kr}\right)\Big|_{\theta=0}^{\pi} k^2\, dk = \frac{4\pi}{r}\int_0^{k_F} k\sin kr\, dk$$

$$= \frac{4\pi}{r}\left(\frac{1}{r^2}\sin kr - \frac{k}{r}\cos kr\right)\Big|_0^{k_F} = \frac{4\pi}{r^3}(\sin k_F r - k_F r \cos k_F r)$$

So $$\rho_\uparrow = \frac{ne}{2}\left(1 - \left(\frac{\frac{4\pi}{r^3}(\sin k_F r - k_F r \cos k_F r)}{\frac{4}{3}\pi k_F^3}\right)^2\right) = \frac{ne}{2}\left(1 - 9\left[\frac{\sin k_F r - k_F r \cos k_F r}{(k_F r)^3}\right]^2\right)$$

$\sin x \sim x - \frac{x^3}{3!}$ and $\cos x \sim 1 - \frac{x^2}{2!}$ so for small r,

$$1 - 9\left(\frac{\sin x - x\cos x}{x^3}\right)^2 \sim 1 - 9\left(\frac{(x - \frac{x^3}{3!}) - x(1 - \frac{x^2}{2!})}{x^3}\right)^2 = 1 - 9\left(\frac{1}{3}\right)^2 = 0$$

→ same-spin electrons are excluded from each other to high order in r!

"Exchange hole"

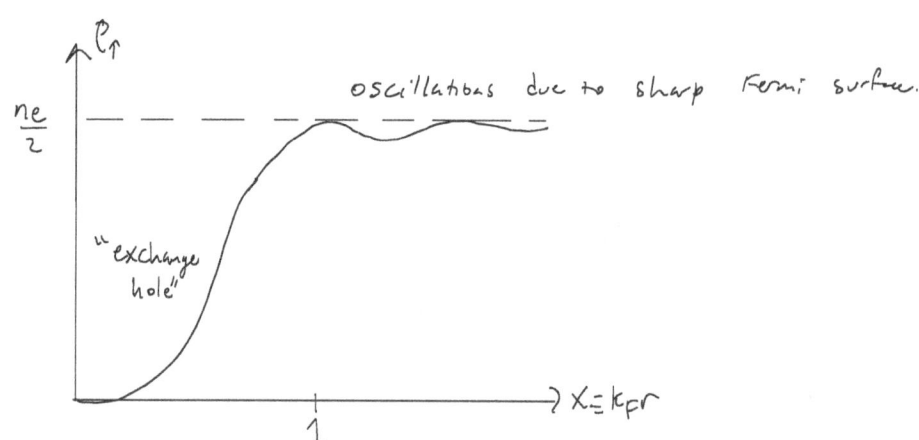

→ exchange hole reduces local charge density + minimizes screening of atomic potentials: lowers system energy. Energy is lowest if all electrons are same spin
→ ferromagnetism is favored!
So we expect nonzero molecular field: Is it enough to drive a phase transition??

Ferromagnetism in the free electron gas

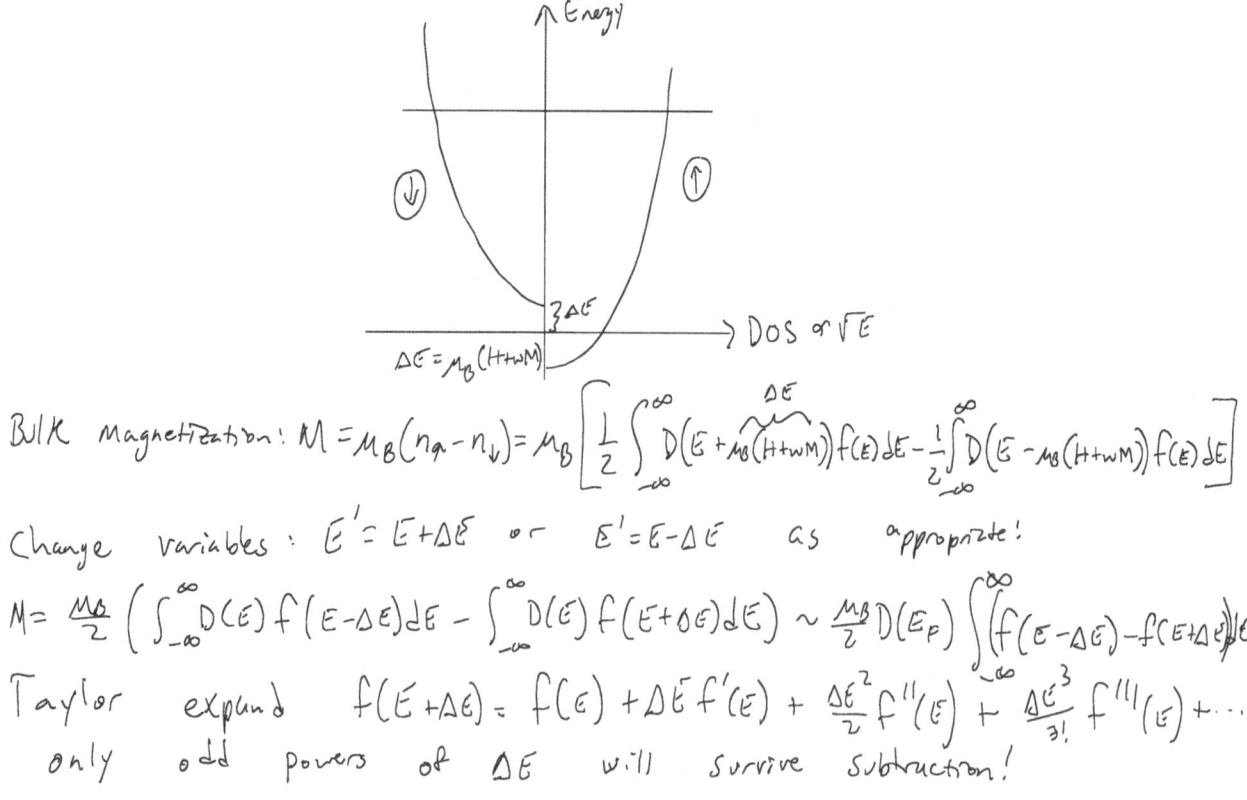

Bulk Magnetization: $M = \mu_B(n_\uparrow - n_\downarrow) = \mu_B \left[\frac{1}{2}\int_{-\infty}^{\infty} D(E + \mu_B(H+wM)) f(E) dE - \frac{1}{2}\int_{-\infty}^{\infty} D(E - \mu_B(H+wM)) f(E) dE \right]$

Change variables: $E' = E + \Delta E$ or $E' = E - \Delta E$ as appropriate!

$M = \frac{\mu_B}{2} \left(\int_{-\infty}^{\infty} D(E) f(E-\Delta E) dE - \int_{-\infty}^{\infty} D(E) f(E+\Delta E) dE \right) \sim \frac{\mu_B}{2} D(E_F) \int_{-\infty}^{\infty} (f(E-\Delta E) - f(E+\Delta E)) dE$

Taylor expand $f(E+\Delta E) = f(E) + \Delta E f'(E) + \frac{\Delta E^2}{2} f''(E) + \frac{\Delta E^3}{3!} f'''(E) + \ldots$
only odd powers of ΔE will survive subtraction!

Stoner criterion for Ferromagnetism

$$M \sim \frac{\mu_B}{2} D(E_F) \int_{-\infty}^{\infty} \left(-2\Delta E f'(E) - \frac{2}{3!} \Delta E^3 f'''(E) \right) dE$$

Note that $f'(E) < 0$ and $f'''(E) > 0$

For $M > 0$ when $H = 0$,

$$M + \mu_B^2 D(E_F) w M \int_{-\infty}^{\infty} f'(E) dE < 0 \longrightarrow 1 - \mu_B^2 D(E_F) w < 0$$

(note: $\int f'(E) dE = -1$)

This puts constraint on molecular field w for Ferromagnetism:

$$w > \frac{1}{\mu_B^2 D(E_F)} \quad \text{(FM favored in metals w/ large } D(E_F) \to \text{large } m^* \text{ d-band)}$$

Even when this condition is not satisfied, Pauli susceptibility becomes enhanced: $M = \mu_B^2 D(E_F)(H + wM) = \chi_0(H + wM) \to M = \frac{\chi_0}{1 - w\chi_0} H$

c.f. "nearly FM" elements Pt, Pd, Rh.

Transition Temperature

Approximate continuous DOS w/ discrete spectrum
c.f. d-electrons in transition metals

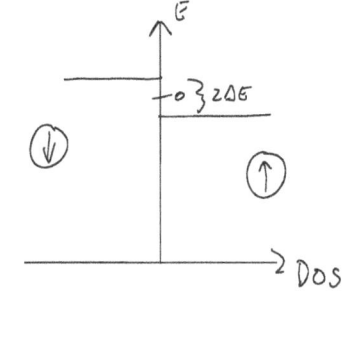

$$M = \mu_B(n_\uparrow - n_\downarrow) = \mu_B \left(\frac{1}{1 + e^{-\mu_B w M / k_B T}} - \frac{1}{1 + e^{+\mu_B w M / k_B T}} \right)$$

$$\frac{M}{\mu_B} = \tanh\left(\frac{\mu_B w M}{2 k_B T}\right) = \tanh\left(\frac{\frac{\mu_B^2 w}{2 k_B}}{T} \frac{M}{\mu_B}\right) = \tanh\left(\frac{T_c}{T} \frac{M}{\mu_B}\right) \equiv \tanh(x)$$

Asymptotic behavior

- For $T \to 0$, $\tanh(x) \to 1$ so $M \to \mu_B$. Approaching $T = 0$,
$$\tanh(x) = \frac{e^x - e^{-x}}{e^x + e^{-x}} = \frac{1 - e^{-2x}}{1 + e^{-2x}} \approx (1 - e^{-2x})^2 \sim 1 - 2e^{-2x} \text{ so } \frac{M}{\mu_B} \sim 1 - 2e^{-2\frac{T_c}{T}\frac{M}{\mu_B}} \sim 1 - 2e^{-2\frac{T_c}{T}}$$

- For $T \to T_c$, $\frac{M}{\mu_B} \sim \tanh\frac{M}{\mu_B}$ so $M \to 0$ and $x \to 0$. Approaching T_c from below,
$$\tanh(x) \sim x - \frac{x^3}{3} \text{ so } \frac{M}{\mu_B} = \frac{T}{T_c} x = x - \frac{x^3}{3} \longrightarrow x \sim \frac{M}{\mu_B} = \sqrt{3}\left(1 - \frac{T}{T_c}\right)^{1/2}$$

Paramagnetic regime (T>T_C)

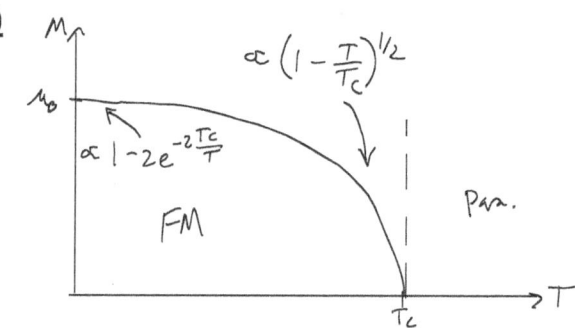

In paramagnetic limit, $T \to \infty$ so X small. Expand Fermi-Dirac

$$\frac{1}{e^{-x}+1} - \frac{1}{e^{x}+1} \approx \frac{1}{2-x} - \frac{1}{2+x} \approx \frac{1}{2}\left(\left(1+\frac{x}{2}\right)-\left(1-\frac{x}{2}\right)\right) = \frac{x}{2}$$

$X \equiv \frac{\mu_B}{k_B T}(H+wM)$, in an external field H, so

$$M = \mu_B\left(\frac{1}{e^{-x}+1}-\frac{1}{e^{x}+1}\right) \approx \frac{\mu_B^2}{2k_B T}(H+wM) \to M \approx \frac{\frac{\mu_B^2}{2k_B T}}{1-\frac{\mu_B^2 w}{2k_B T}}H = \frac{C}{T-T_C}H$$

This is the "Curie-Weiss Law" from the local moment model!

Ferromagnetic domains

System can minimize total energy, eliminating magnetostatic stray field by fracturing into magnetic domains. This process is a competition between magnetostatic energy and the exchange energy associated w/ wall formation

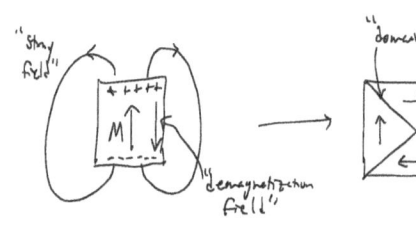

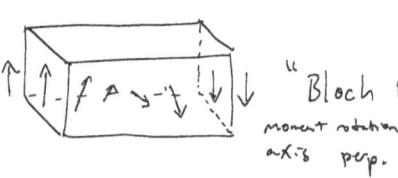

"Bloch Wall" moment rotation along axis perp. to wall

domain wall energy

Nearest-neighbor exchange $E = -\beta M_1 M_2 \cos\theta \approx -\beta M_1 M_2 \left(1-\frac{\delta\theta^2}{2}\right) = C + \frac{\beta M^2}{2}\delta\theta^2$

Since we can redefine zero energy, and $\delta\theta = \frac{\pi}{N}$ (wall is N moments thick)

Energy density $u = \frac{\beta M^2}{2a^3}\left(\frac{\pi}{N}\right)^2$ ($a \equiv$ lattice const, spacing between N.N.'s)

Areal density is $u \cdot Na = \frac{\beta M^2}{2a^3}\left(\frac{\pi}{N}\right)^2 \cdot Na = A\frac{\pi^2}{aN}$ ($A \equiv \frac{\beta M^2}{2a}$ "exchange constant")

This favors wide walls (large N)

Crystal anisotropy

Spin-orbit interaction tends to align moments along crystal axis, so we must pay an energetic price to have moments in the wall oriented at intermediate directions. This obviously favors thin domain walls. Total energy is:

$$E_{wall} = \underbrace{K}_{\text{anisotropy const}} N a + \underbrace{A}_{\text{exchange}} \frac{\pi^2}{Na}$$

Minimize with $\frac{dE_{wall}}{dN} = 0 \rightarrow Ka - A\frac{\pi^2}{N^2 a} = 0 \rightarrow Na = \sqrt{A \frac{\pi^2}{K}}$

Example: $A \sim 10^6 \, eV/cm$ and $K \sim \frac{10^{18} eV}{cm^3}$ so $Na \sim 10^{-6} cm$ (10 nm)
FMs of this size cannot form multiple domains $\rightarrow$ "single domain FM"

Magnetization of FM

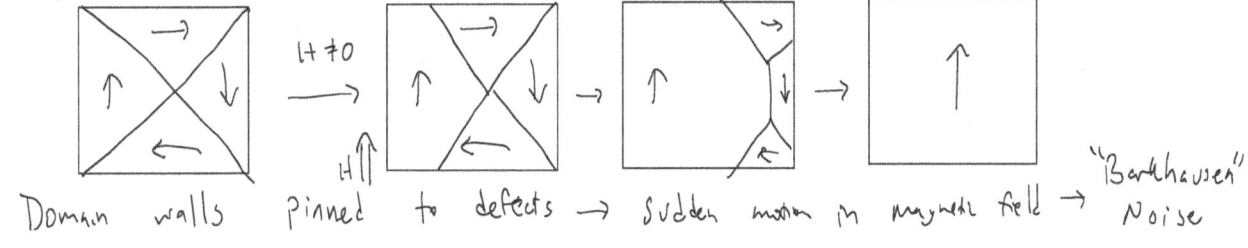

Domain walls pinned to defects $\rightarrow$ sudden motion in magnetic field $\rightarrow$ "Barkhausen" noise

Coherent rotation   "Stoner-Wohlfarth" Model

The energy associated with the interaction of the magnetization with the demagnetizing field yields a geometric "shape anisotropy".

$$E = \underbrace{\tfrac{1}{2}M^2 \left(D_b \cos^2(\phi-\theta) + D_a \sin^2(\phi-\theta) \right)}_{\text{Uniaxial anisotropy}} - \underbrace{HM\cos\phi}_{\text{Zeeman}}$$

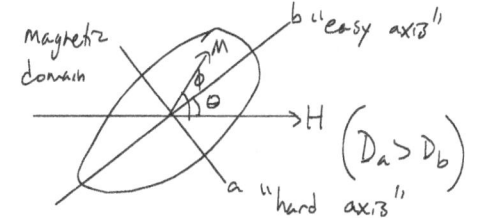

$(D_a > D_b)$

Using $\cos^2 x = \tfrac{1}{2}(1+\cos 2x)$ and $\sin^2 x = \tfrac{1}{2}(1-\cos 2x)$,

$$E = \tfrac{1}{4}M^2(D_a+D_b) - \tfrac{1}{4}M^2(D_a-D_b)\cos(2(\phi-\theta)) - HM\cos\phi$$

Find equilibrium orientation of magnetization ϕ by setting

$$\frac{dE}{d\phi} = \tfrac{1}{2}M^2(D_a-D_b)\sin(2(\phi-\theta)) + MH\sin\phi = 0 \qquad \text{(special cases follow)}$$

For θ=0 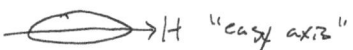 "easy axis"

$$\frac{1}{2}\sin 2\phi + h\sin\phi = 0$$

$\left(h \equiv \dfrac{H}{(D_a - D_b)M}\right.$ "reduced field"$\left.\right)$

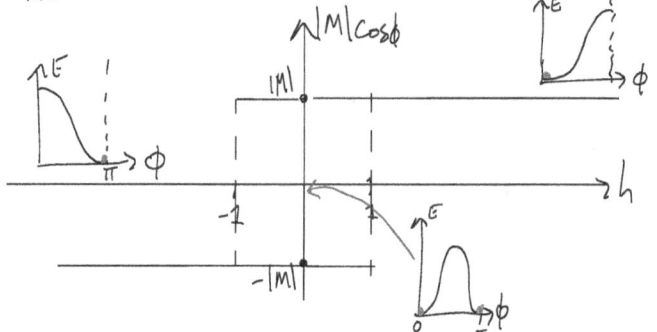

For θ=π/2 →H "hard axis"

$$\frac{1}{2}\sin\left(2\left(\phi - \frac{\pi}{2}\right)\right) + h\sin\phi = 0$$

$$-\frac{1}{2}\sin 2\phi + h\sin\phi = 0$$

$$\cancel{\sin\phi}\cos\phi = h\cancel{\sin\phi}$$

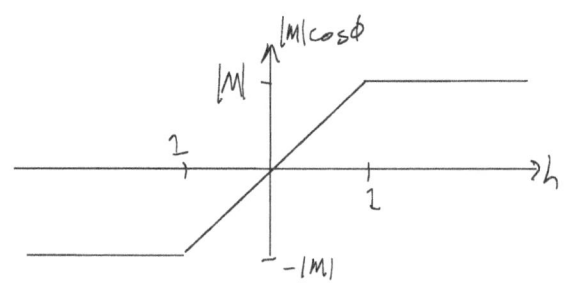

For θ=π/4  →h

$$\frac{1}{2}\sin\left(2\left(\phi - \frac{\pi}{4}\right)\right) + h\sin\phi = 0$$

$$\frac{1}{2}\sin\left(2\phi - \frac{\pi}{2}\right) + h\sin\phi = 0$$

$$-\frac{1}{2}\cos 2\phi = -h\sin\phi$$

If $h = \frac{1}{2}$, $\phi = \frac{\pi}{2} \rightarrow \cos\phi = 0$

$h = -\frac{1}{2}$, $\phi = \frac{3\pi}{2} \rightarrow \cos\phi = 0$

$h = 0$, $\phi = \pm\frac{\pi}{4} \rightarrow \cos\phi = \pm\frac{\sqrt{2}}{2}$

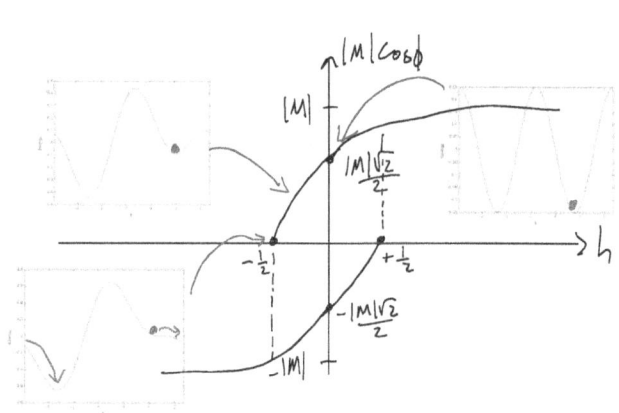

Ensemble average over all θ

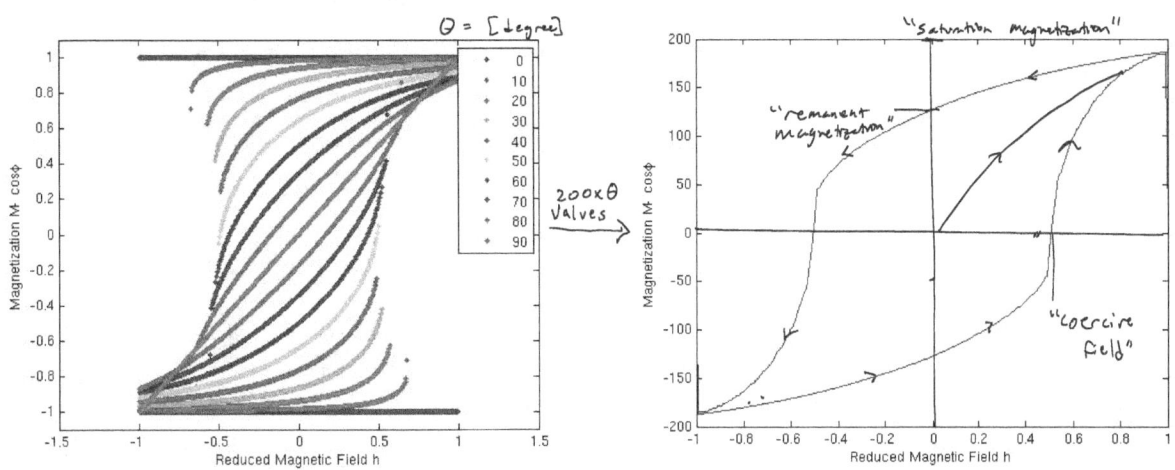

To magnetize a randomly-oriented ensemble w/ M=0, just increase H past coercive field. To demagnetize sample, just oscillate H from large amplitude to small. As we pass to lower field amplitude, the coercive field is not exceeded and the domain is stuck in random orientation!

Applications of Ferromagnetism   Remanence ≡ "Memory"

Information is encoded into the $+M="1"$ and $-M="0"$ states at $H=0$. But, one must consider long-term stability against thermal fluctuations: Stability of magnetic state determined by Arrhenius' Law $t = t_o e^{E_a/k_B T}$ ($t_o \sim 1 ns$)

For $t = 10 yrs$, then $E_a \cong 40 k_B T = KV$ (K = anisotropy const, V = volume). If volume small enough, $E_a \sim k_B T$ and "superparamagnetism" results → magnetic fluctuation pinned by external field (not anisotropy) so remanence (and hence memory) is lost.

This limits conventional information density on hard drive to $\sim 100 Gb/in$. Increasing K can increase this limit, but then the coercive field rises so it becomes harder to "write" information!

Tunnel Magnetoresistance (TMR)

Sensitive magnetic field sensors are required to "read" information encoded in magnetic state of a FM domain. TMR is the current technology in the "read head" floating above a spinning hard drive platter. The stray field from a "bit" on the platter manipulates the magnetization state of a "free" FM layer relative to a "fixed" layer, causing resistance change for transport of electrons from one (cathode) to the other (anode).

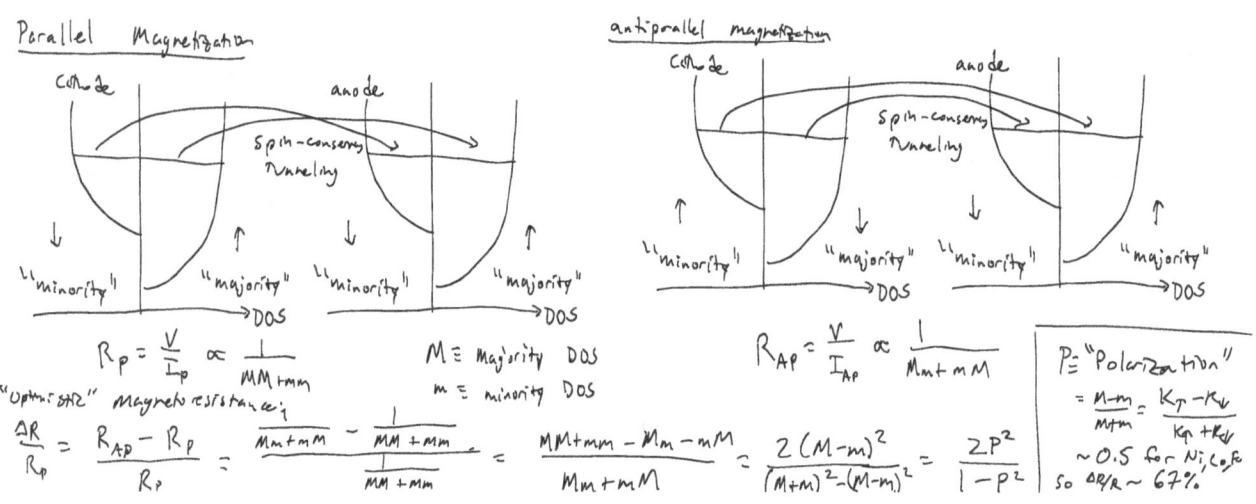

$R_P = \frac{V}{I_P} \propto \frac{1}{MM + mm}$

$M \equiv$ Majority DOS
$m \equiv$ minority DOS

$R_{AP} = \frac{V}{I_{AP}} \propto \frac{1}{Mm + mM}$

$P \equiv$ "Polarization"
$= \frac{M-m}{M+m} = \frac{K_\uparrow - K_\downarrow}{K_\uparrow + K_\downarrow}$
~ 0.5 for Ni, Co, Fe
so $\Delta R/R \sim 67\%$

"Optimized" Magnetoresistance:

$\frac{\Delta R}{R_P} = \frac{R_{AP} - R_P}{R_P} = \frac{\frac{1}{Mm+mM} - \frac{1}{MM+mm}}{\frac{1}{MM+mm}} = \frac{MM+mm - Mm - mM}{Mm+mM} = \frac{2(M-m)^2}{(M+m)^2 - (M-m)^2} = \frac{2P^2}{1-P^2}$

Tunnel Polarization

We considered only spin asymmetry caused by initial and final DOS, but did not consider the tunnel process in determining transport asymmetry. Transmission coefficient can be approximated by incoherent product of interfacial transmission coefs:

$T \sim \underbrace{\frac{4 K_c \kappa}{K_c^2 + \kappa^2}}_{\text{cathode-barrier}} \underbrace{\frac{4 K_a \kappa}{K_a^2 + \kappa^2}}_{\text{barrier-anode}}$

where $K_c = K_\uparrow$ or $K_\downarrow$

tunnel spin polarization:

$P = \frac{T_\uparrow - T_\downarrow}{T_\uparrow + T_\downarrow} = \frac{\frac{K_\uparrow}{K_\uparrow^2 + \kappa^2} - \frac{K_\downarrow}{K_\downarrow^2 + \kappa^2}}{\frac{K_\uparrow}{K_\uparrow^2 + \kappa^2} + \frac{K_\downarrow}{K_\downarrow^2 + \kappa^2}} = \frac{K_\uparrow(K_\downarrow^2 + \kappa^2) - K_\downarrow(K_\uparrow^2 + \kappa^2)}{K_\uparrow(K_\downarrow^2 + \kappa^2) + K_\downarrow(K_\uparrow^2 + \kappa^2)}$

$= \frac{K_\uparrow - K_\downarrow}{K_\uparrow + K_\downarrow} \cdot \frac{\kappa^2 - K_\uparrow K_\downarrow}{\kappa^2 + K_\uparrow K_\downarrow}$ "Slonczewski Polarization" (c.f. $P = \frac{K_\uparrow - K_\downarrow}{K_\uparrow + K_\downarrow}$ above)

This tunnel-dependent polarization is especially important when considering spin transport (in addition to charge) from a FM to a normal conductor. First, we need to generalize Ohm's law $J = \sigma E$.

Generalization of Ohm's Law: inclusion of diffusion current

Electric field drives a "drift" current proportional to conductivity σ, but concentration gradient drives a "diffusion" current proportional to Diffusion coef. D:

$$J = \overbrace{\sigma \mathcal{E}}^{\text{drift}} - \overbrace{eD\nabla n}^{\text{diffusion}} = \sigma\left(\nabla(-\phi) - \frac{eD}{\sigma}\nabla n\right) = \sigma\left(\nabla(-\phi) - \frac{eD}{ne\mu}\nabla n\right) = \sigma\nabla(-\tilde{\phi})$$

(diffusion const.)

$\tilde{\phi} \equiv \phi + \frac{D}{\mu}\ln\frac{n}{n_0}$ "electrochemical potential" (Note, $D/\mu = \frac{k_BT}{e}$ for nondegenerate electron gas)

At an interface with $\sigma_1 \neq \sigma_2$:

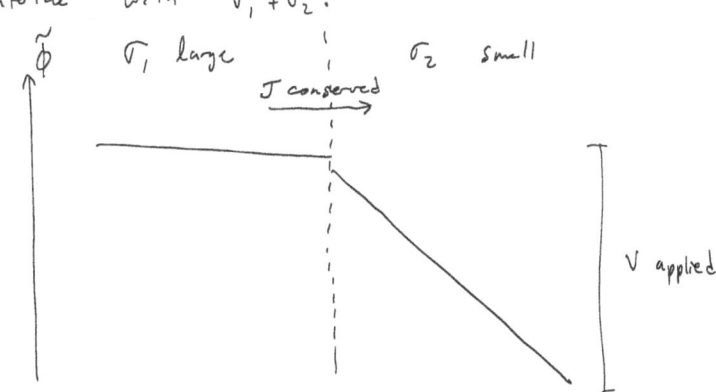

Spin-dependent Ohmic transport

$J_\uparrow = \sigma_\uparrow(-\nabla\tilde{\phi}_\uparrow)$, $J_\downarrow = \sigma_\downarrow(-\nabla\tilde{\phi}_\downarrow)$

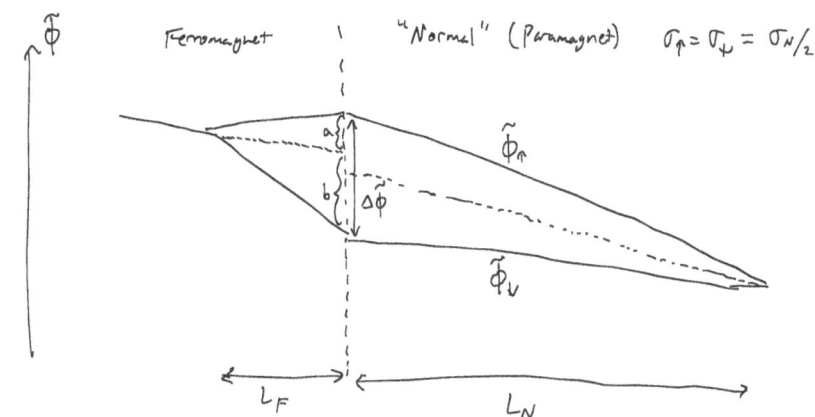

We want Polarization in normal metal $P = \frac{J_\uparrow - J_\downarrow}{J_\uparrow + J_\downarrow} = \frac{J_\uparrow - J_\downarrow}{J} \neq 0$ so $\nabla\tilde{\phi}_\uparrow \neq \nabla\tilde{\phi}_\downarrow$!

This causes splitting $\Delta\tilde{\phi}$ at interface.

Note: $J_\uparrow = J\left(\frac{1+P}{2}\right)$ $J_\downarrow = J\left(\frac{1-P}{2}\right)$

Likewise, in FM we have bulk polarization $\beta = \frac{\sigma_\uparrow - \sigma_\downarrow}{\sigma_\uparrow + \sigma_\downarrow}$

so that $\sigma_\uparrow = \sigma_F\left(\frac{1+\beta}{2}\right)$, $\sigma_\downarrow = \sigma_F\left(\frac{1-\beta}{2}\right)$ where $\sigma_F = \sigma_\uparrow + \sigma_\downarrow$

Can P=β?

Normal side: $J_\uparrow = \frac{J}{2} + \frac{\sigma_N}{2}\frac{\Delta\tilde{\phi}_N}{L_N}$ and $J_\downarrow = \frac{J}{2} - \frac{\sigma_N}{2}\frac{\Delta\tilde{\phi}_N}{L_N}$ so $P = \frac{\sigma_N \Delta\tilde{\phi}_N}{2L_N J}$

But what is $\Delta\tilde{\phi}_N$?

FM side: $J_\uparrow = \sigma_\uparrow\left[\frac{J}{\sigma_F} - \frac{a}{L_F}\right]$ and $J_\downarrow = \sigma_\downarrow\left[\frac{J}{\sigma_F} + \frac{b}{L_F}\right]$

So $\Delta\tilde{\phi}_F = a+b = L_F\left[\left(\frac{J}{\sigma_F} - \frac{J_\uparrow}{\sigma_\uparrow}\right) + \left(\frac{J_\downarrow}{\sigma_\downarrow} - \frac{J}{\sigma_F}\right)\right] = L_F\left[\frac{J_\downarrow}{\sigma_\downarrow} - \frac{J_\uparrow}{\sigma_\uparrow}\right]$

$= L_F\left[\frac{J(\frac{1-P}{2})}{\sigma_F(\frac{1-\beta}{2})} - \frac{J(\frac{1+P}{2})}{\sigma_F(\frac{1+\beta}{2})}\right] = J\frac{L_F}{\sigma_F}\frac{(1-P)(1+\beta) - (1+P)(1-\beta)}{1-\beta^2} = J\frac{L_F}{\sigma_F}\frac{\beta-P}{1-\beta^2}$

Equating $\Delta\tilde{\phi}_F = \Delta\tilde{\phi}_N$,

$\frac{2L_N P}{\sigma_N}\cancel{J} = \cancel{J}\frac{L_F}{\sigma_F}\frac{\beta-P}{1-\beta^2} \longrightarrow P = \frac{\beta\eta}{1+\eta-\beta^2}$ where $\eta = \frac{L_F \sigma_N}{\sigma_F L_N}$

$\Rightarrow$ only if $\beta = 1$ (perfect "half metallic" FM) does $P = \beta$ independent of η.
But, $\beta \sim 0.5$ for Ni, Fe, Co. Then η dominates; unfortunately, $L_F \ll L_N$ for typical metals, and (especially if the nonmagnet is a semiconductor), $\sigma_F \gg \sigma_N$ so η and hence P is small!

Relaxation of electrochemical continuity

The problem is driven by $\Delta\tilde{\phi}$ on the FM side, opposing flow of spin up!
The solution is to minimize the splitting w/ intermediate layer.

Using expressions previously derived, ($G \equiv$ barrier conductance)

$\Delta\tilde{\phi}_N - \Delta\tilde{\phi}_F = \left(\frac{J_\downarrow}{G_\downarrow} - \frac{J_\uparrow}{G_\uparrow}\right)$

$\Longrightarrow P = \dfrac{\beta\eta\left(1 + \frac{1}{4\beta}\frac{\sigma_F}{L_F}\left(\frac{1}{G_\downarrow} - \frac{1}{G_\uparrow}\right)(1-\beta^2)\right)}{1+\eta-\beta^2 + \left(\frac{\sigma_N}{4L_N}\left(\frac{1}{G_\downarrow} + \frac{1}{G_\uparrow}\right)(1-\beta^2)\right)}$

In low G limit, $P \approx \frac{G_\uparrow - G_\downarrow}{G_\uparrow + G_\downarrow}$, so an ohmic, nonmagnetic intermediate layer (with $G_\uparrow = G_\downarrow$) only makes the problem worse!

Tunneling (not ohmic transport) gives $G_\uparrow \neq G_\downarrow$ because $G \propto$ Transmission. Therefore Slonczewski's result saves the day for spin injection!

Electrons in 1-d periodic potential

We still haven't explained the existence of crystalline <u>insulators</u>. To do so, we must account for the periodic electrostatic potential from the atomic ions in the lattice.

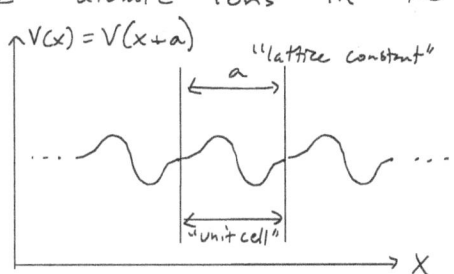

$V(x)$ can be expanded in a Fourier series:

$$V(x) = \sum_G V_G e^{iGx}, \quad G = G_h = h\frac{2\pi}{a} = 0, \pm\frac{2\pi}{a}, \pm\frac{4\pi}{a}, \text{ etc.} \quad (h = \ldots, -1, 0, 1, \ldots)$$

G's are discrete-valued "reciprocal lattice numbers" satisfying periodic symmetry.

We want to solve Schrödinger eqn:

$$\left[-\frac{\hbar^2}{2m}\frac{d^2}{dx^2} + V(x)\right]\psi = E\psi \quad \text{In general, sol'ns not plane waves!}$$

Bloch's theorem

Solution: $\psi = e^{ikx} u(x)$ "Bloch wave" where $u(x)$ has the same periodic symmetry as the crystal lattice.

Therefore, $u(x) = \sum_G C_G e^{iGx}$

Substitute:

$$-\frac{\hbar^2}{2m}\left(e^{ikx} u(x)\right)'' + V(x) e^{ikx} u(x) = E e^{ikx} u(x)$$

$$-\frac{\hbar^2}{2m}\left(u'' + 2iku' - k^2 u\right)e^{ikx} + V e^{ikx} u = E e^{ikx} u$$

$$-\frac{\hbar^2}{2m}\left(\frac{d}{dx} + ik\right)^2 u + V u = E u$$

This is a "Schrödinger eqn" for the envelope fn. $u(x)$.

59

k-dot-p method

Since momentum $p = \frac{\hbar}{i}\frac{d}{dx}$, we can write our "Schrödinger eqn" as

$$\frac{(p+\hbar k)^2}{2m} u + V u = E u$$

$$\left[\underbrace{\frac{p^2}{2m} + V}_{H_0} + \underbrace{\frac{\hbar^2 k^2}{2m} + \frac{\hbar}{m} k \cdot p}_{H_1} \right] u = E u$$

This suggests a perturbative expansion about $k=0$. More in this direction later...

Fourier representation   (Substitute $u(x) = \sum_G C_G e^{iGx}$)

In this representation, $\frac{d}{dx} \to iG$ so $-\frac{\hbar^2}{2m}\left(\frac{d}{dx} + ik\right)^2 u + V u = E u$ becomes

$$+\frac{\hbar^2}{2m} \sum_G (k+G)^2 C_G e^{iGx} + \sum_{G'} V_{G'} e^{iG'x} \sum_G C_G e^{iGx} = E \sum_G C_G e^{iGx}$$

Simplest case $V_G \to 0$ for all G ($V(x) = 0$, but retain periodic symmetry)

Since e^{iGx} are orthonormal, all like terms must be equal, so we have $E = \frac{\hbar^2}{2m}(k+G)^2$. This is identical to the case of continuous translational symmetry, except $k \to k+G$!

60

Free electron bandstructure

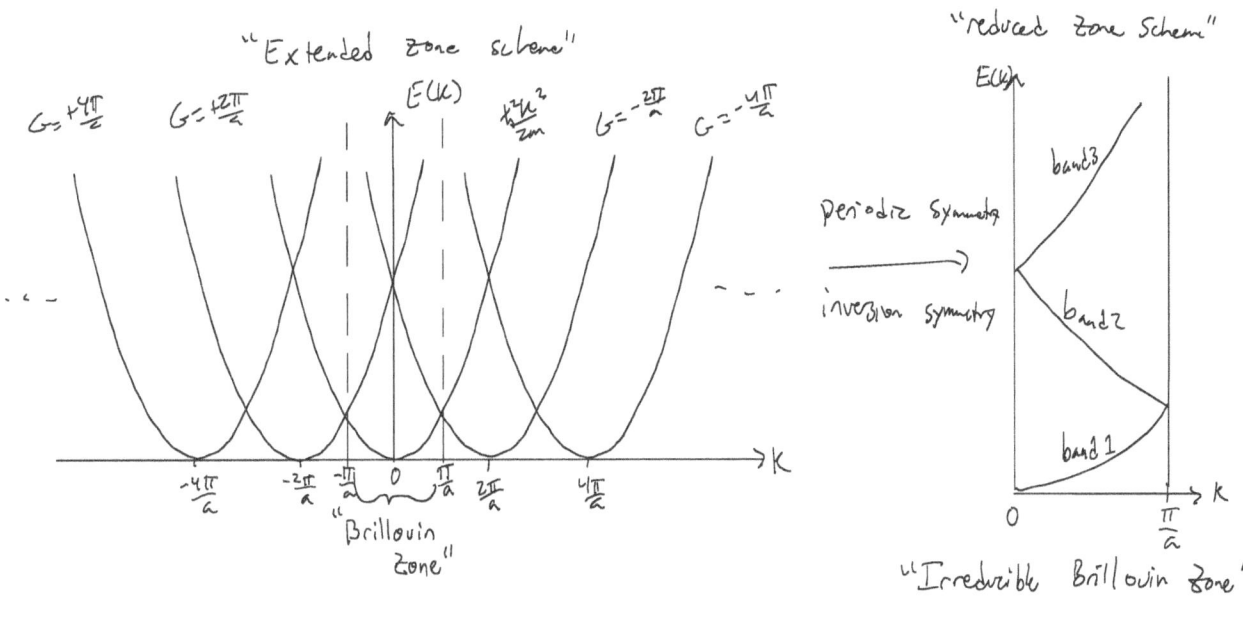

But this <u>still</u> describes a metal! must understand $V \neq 0$ case!

Nonzero potential

Potential energy term in Schrodinger eqn is

$$\sum_{G'} V_{G'} e^{iG'x} \sum_{G} C_G e^{iGx} = \sum_{h'} V_{h'} e^{iG_{h'}x} \sum_{h} C_h e^{iG_h x} = \sum_{h}\sum_{h'} V_{h'} C_h e^{i(G_h+G_{h'})x}$$

$$= \sum_{h}\sum_{h'} V_{h'} C_h e^{iG_{h+h'}x}$$

Variable substitution $h+h' \to h$: $\quad \sum_{h}\sum_{h'} V_{h'} C_{h-h'} e^{iG_h x}$

$$\sum_{h} \left[\frac{\hbar^2}{2m}(k+G_h)^2 C_h + \sum_{h'} V_{h'} C_{h-h'} \right] e^{iG_h x} = \sum_{h} E C_h e^{iG_h x}$$

Since $e^{iG_h x}$ basis fns are an orthonormal set, each coeff. must separately satisfy

$$\frac{\hbar^2}{2m}(k+G_h)^2 C_h + \sum_{h'} V_{h'} C_{h-h'} = E C_h \qquad (\text{for all } h!)$$

Infinite matrix eigenvalue problem

$(h=-1)$ $\ldots + V_1 C_{-2} + \frac{\hbar^2}{2m}(k+G_{-1})^2 C_{-1} + V_0 C_{-1} + V_{-1} C_0 + V_{-2} C_1 + \ldots = E C_{-1}$

$(h=0)$ $\ldots + V_2 C_{-2} + V_1 C_{-1} + \frac{\hbar^2}{2m}(k+G_0)^2 C_0 + V_0 C_0 + V_{-1} C_1 + \ldots = E C_0$

$(h=+1)$ $\ldots + V_3 C_{-2} + V_2 C_{-1} + V_1 C_0 + \frac{\hbar^2}{2m}(k+G_1)^2 C_1 + V_0 C_1 + \ldots = E C_1$

This infinite system of linear equations is equivalent to:

$$\begin{bmatrix} \vdots & & & & & \\ \cdots & V_2 & V_1 & \frac{\hbar^2}{2m}(k+G_0)^2+V_0 & V_{-1} & V_{-2} \cdots \\ \cdots & & V_2 & V_1 & \frac{\hbar^2}{2m}(k+G_1)^2+V_0 & V_{-1} \cdots \\ & & & & \vdots & \end{bmatrix} \begin{bmatrix} \vdots \\ C_{-1} \\ C_0 \\ C_{+1} \\ \vdots \end{bmatrix} = E \begin{bmatrix} \vdots \\ C_{-1} \\ C_0 \\ C_{+1} \\ \vdots \end{bmatrix}$$

An infinite matrix eigenvalue problem!

$\mathcal{H}\psi = E\psi$

Since $V(x)$ is real, $V_G = V_{-G}^*$ so $\mathcal{H}$ is Hermitian, w/ real eigenvalues

Finite matrix eigenvalue problem

But we can't solve for eigenvalues/eigenvectors of an infinite matrix. Solution:

Cutoff values of G to finite number $\rightarrow$ equivalent to finite spatial resolution. $x \rightarrow n\frac{a}{N}$, n an integer, $1 < n < N$, N is total # of Gs.

$e^{i(G_{h'} + G_h)x} = e^{i(h+h')\frac{2\pi}{a}\frac{na}{N}}$ $\rightarrow$ periodic in $h+h'$ w/ period N

This modular counting results in cyclic permutations of $V_{h'}$!

Example: $N=3$, $h = -1, 0, +1$

$(h=-1)$
$(h=0)$
$(h=+1)$

$$\begin{pmatrix} \frac{\hbar^2}{2m}(k+G_{-1})^2+V_0 & V_{-1} & V_{+1} \\ V_{+1} & \frac{\hbar^2}{2m}(k+G_0)^2+V_0 & V_{-1} \\ V_{-1} & V_{+1} & \frac{\hbar^2}{2m}(k+G_1)^2+V_0 \end{pmatrix} \begin{bmatrix} C_{-1} \\ C_0 \\ C_{+1} \end{bmatrix} = E \begin{bmatrix} C_{-1} \\ C_0 \\ C_{+1} \end{bmatrix}$$

Numerical diagonalization

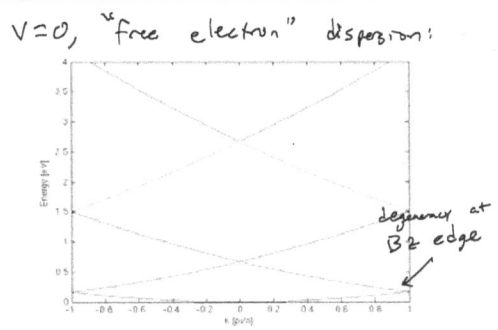

$V=0$, "free electron" dispersion:

degeneracy at BZ edge

The broken continuous translational symmetry induced by $V(x)$ consequently breaks band degeneracy at BZ edge. This results in "bandgaps", which can be further understood by inspecting the alignment of probability density $\psi^*\psi$ w.r.t. the potential $V(x)$ for k at the BZ edge.

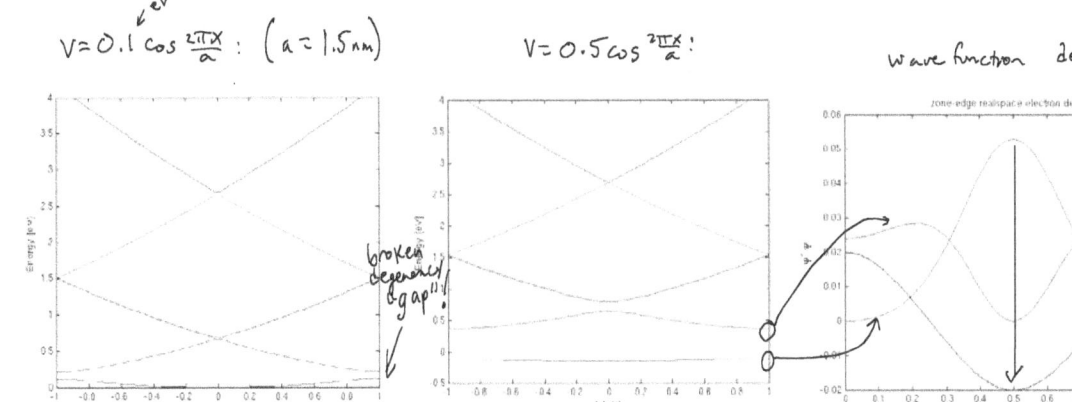

$V = 0.1 \cos\frac{2\pi x}{a}$ eV : $(a = 1.5\,nm)$

$V = 0.5 \cos\frac{2\pi x}{a}$:

broken degeneracy & gap!

wave function density $\psi^*\psi$

Bandgaps

Why are bandgaps induced by nonzero potential giving off-diagonal elements of the Hamiltonian matrix?

$$\mathcal{H} = \begin{bmatrix} E_1 & \Delta \\ \Delta^* & E_2 \end{bmatrix}$$

At BZ edge, $E_1 = E_2$ so $\mathcal{H} = E_{1,2}\mathbb{1} + \begin{bmatrix} 0 & \Delta \\ \Delta^* & 0 \end{bmatrix}$

eigenvalues are $E_{1,2}$ + roots of $\{E^2 - |\Delta|^2 = 0\}$

$\rightarrow E = E_{1,2} \pm |\Delta| \rightarrow$ a gap of $2|\Delta|$!

This bandgap is essential to explain the existence of electrical insulators: If E_F lies in the gap @ $T=0$, no available states at infinitesimal excitation energies, so asymmetric occupation distributions are impossible! Despite high density of electrons, conductivity $\sigma = 0$.

Also explains optical properties (long wavelength transparency) and photoconductivity $\longrightarrow$ particle detectors!

Fermi energy

Fermi energy is determined by accounting for every electron in unit cell, 2 electrons per band. If the last filled band is just below a bandgap, the material is an "insulator" (or semiconductor as explained later). Otherwise, half-filled bands give "metals" and zero gap gives "semimetal".

Examples:

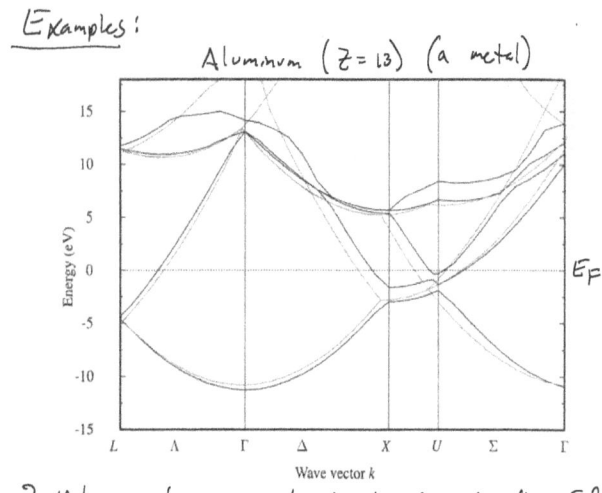

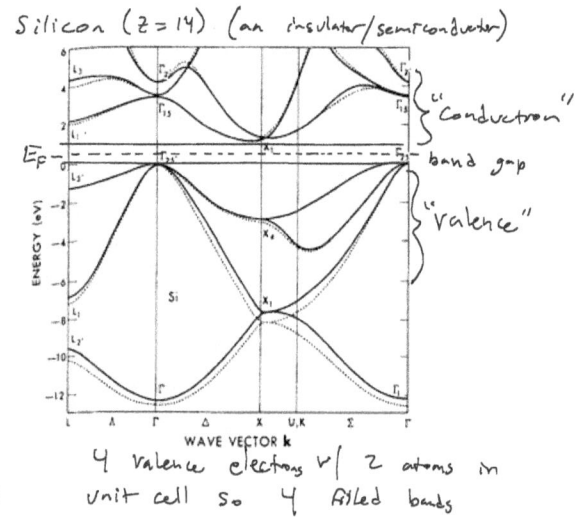

3 valence electrons w/ 1 atom in unit cell! 1.5 filled bands

4 valence electrons w/ 2 atoms in unit cell so 4 filled bands

Bandstructure beyond one dimension

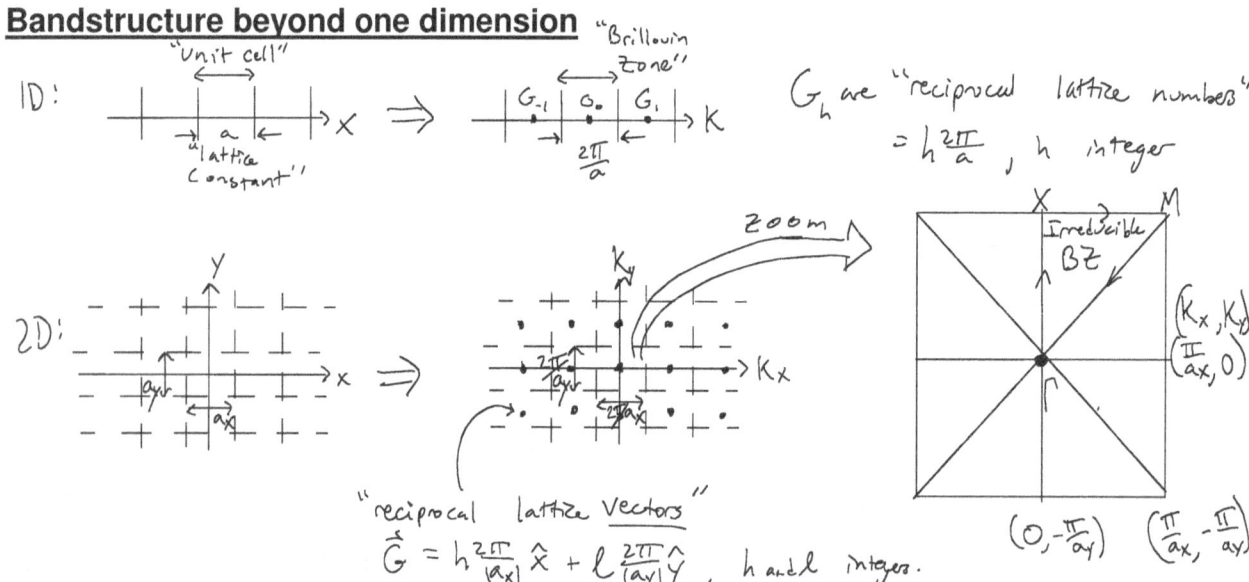

The BZ can be reduced to the IBZ by exploiting inversion + rotation symmetry of k-space reciprocal lattice. Bandstructure is calculated on the perimeter of this IBZ, where points of high symmetry are labeled by Roman letters on BZ boundary, and Greek at interior.

2D square lattice: "nearly-free-electron" bandstructure

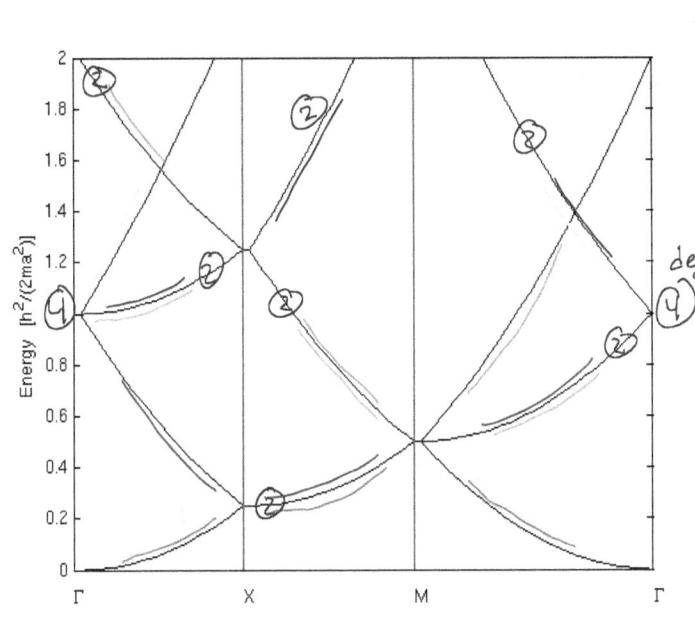

A paraboloid $\frac{\hbar^2}{2m}|\vec{K}-\vec{G}|^2$ sits on every reciprocal lattice point $\vec{G}$ and crosses our path along the IBZ perimeter

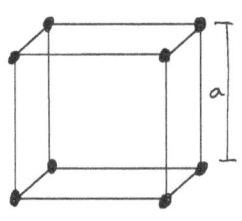

K-space

By symmetry, bands must cross at a degeneracy or have zero slope due to an induced gap!

3 dimensions: primitive unit cell

primitive = includes only 1 lattice point. For example, a primitive unit cell for the "simple cubic" lattice might be chosen w/ the "conventional" unit cell. →

Each lattice point is shared w/ 8 unit cells so this contains $8 \times 1/8 = 1$ total lattice points. However, it is more convenient to choose the Wigner-Seitz unit cell w/ one lattice point at the center so that constructing the BZ is straightforward:

real (x) space
Wigner-Seitz cell:

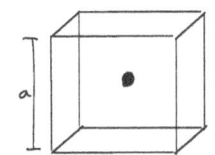

reciprocal (K) space
Brillouin Zone:

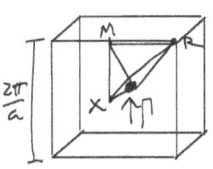

Irreducible BZ (here, 1/48th of BZ volume) is an irregular polyhedron defined by lattice symmetries

related lattices (conventional unit cells):

body-centered cubic (bcc)
(eg. alkali metals)

face-centered cubic (fcc)
(eg. Cu, Ag, Au)

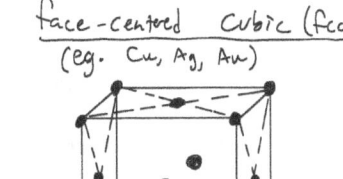
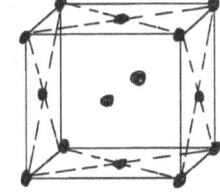

BZs in 3d: bcc and fcc

Generate *primitive* reciprocal lattice vectors from real-space *primitive* lattice vectors via

$$\vec{g}_i \cdot \vec{a}_j = 2\pi \delta_{ij} \implies \vec{g}_1 = 2\pi \frac{\vec{a}_2 \times \vec{a}_3}{\vec{a}_1 \cdot (\vec{a}_2 \times \vec{a}_3)}$$

and cyclic permutations of subscripts

real-space lattice — reciprocal lattice

bcc: "Wigner-Seitz" primitive cell (planes halfway between origin + nearest lattice point)

fcc: (primitive cell contains 1 lattice point)

Note that bcc BZ in k-space is same as fcc W-S cell in x-space, and vice-versa!

IBZs and symmetry points of cubic lattices

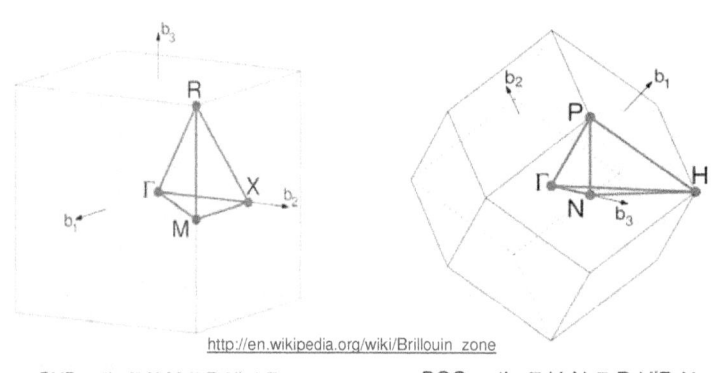
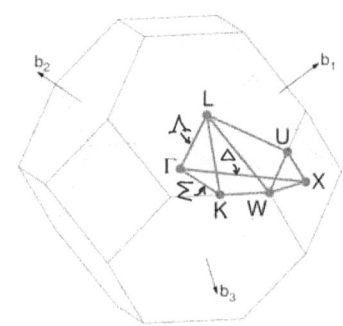

CUB path: Γ-X-M-Γ-R-X|M-R

BCC path: Γ-H-N-Γ-P-H|P-N

FCC path: Γ-X-W-K-Γ-L-U-W-L-K|U-X

[Setyawan & Curtarolo, DOI: 10.1016/j.commatsci.2010.05.010]

These are examples of "Bravais lattices": All lattice points are equivalent. Note, however, that NOT ALL lattices are Bravais!

Example: diamond (two interpenetrating fcc lattices)

Diamond lattice: III-V and IV semiconductors

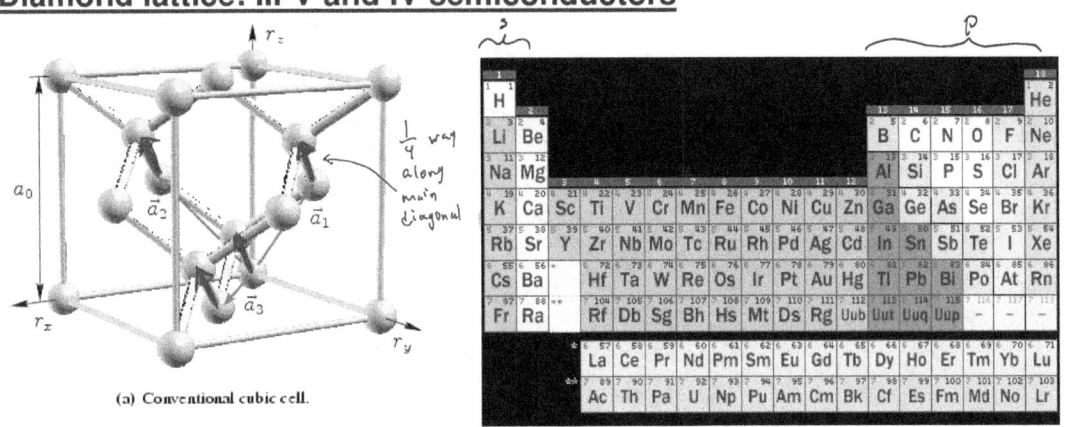

(a) Conventional cubic cell.

Each atom in group IV (C, Si, Ge) has 4× equivalent sp^3 hybridized bonds equally distributed in space. This tetragonal coordination leads to diamond lattice: 2 interpenetrating fcc lattices, separated by 1/4 distance along main diagonal. Zincblende is formed by putting inequivalent atoms (e.g. from groups III and V or II and VI) on the fcc lattices. The BZ is identical to fcc BZ since it is the underlying Bravais lattice (with basis $\vec{d} = 0$ and $\frac{a}{4}(\hat{x}+\hat{y}+\hat{z})$).

Fermi surface

Example: Copper (FCC)

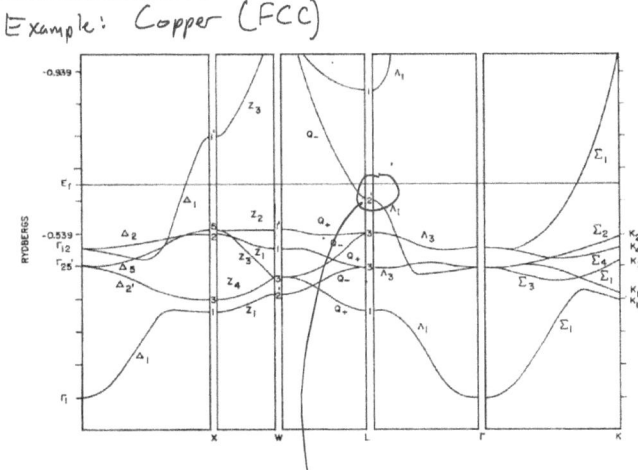

The nearly-free electron bands always meet in degeneracies at the BZ boundary, so the Fermi surface is a sphere (as in the Sommerfeld model neglecting periodicity).

For bands where potential $V(\vec{r})$ has been taken into account, this is obviously not the case due to off-diagonal matrix elements and degeneracy splitting.

For Cu, we have valence $[Ar]\,4s^1\,3d^{10}$, so with 11 electrons, we fill $5\text{-}\frac{1}{2}$ bands to find the Fermi energy. This line intersects a partially filled band along Γ–X, but NOT along Γ–L (111 dir). The resulting Fermi surface merges with another sphere in an adjacent BZ, forming a "neck" that can be seen in deHaas-van Alphen w/ H along (111). Two frequencies in 1/H from "neck" and "belly" result.

Types of bandgaps

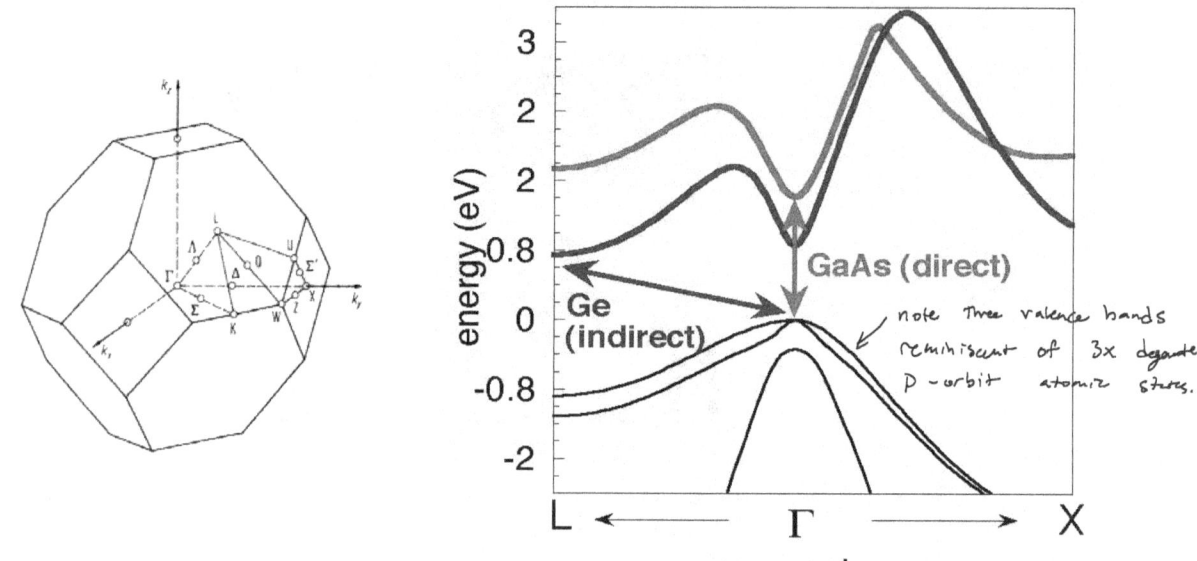

Although many small-gap insulators (semiconductors) have diamond/zincblende lattice, the bandstructure can be very different: varying bandgaps AND varying locations in k-space where conduction band is lowest. This defines two kinds of bandgaps: direct (e.g. GaAs, InP) and indirect (e.g. Si, Ge, GaP).

Constant energy surfaces

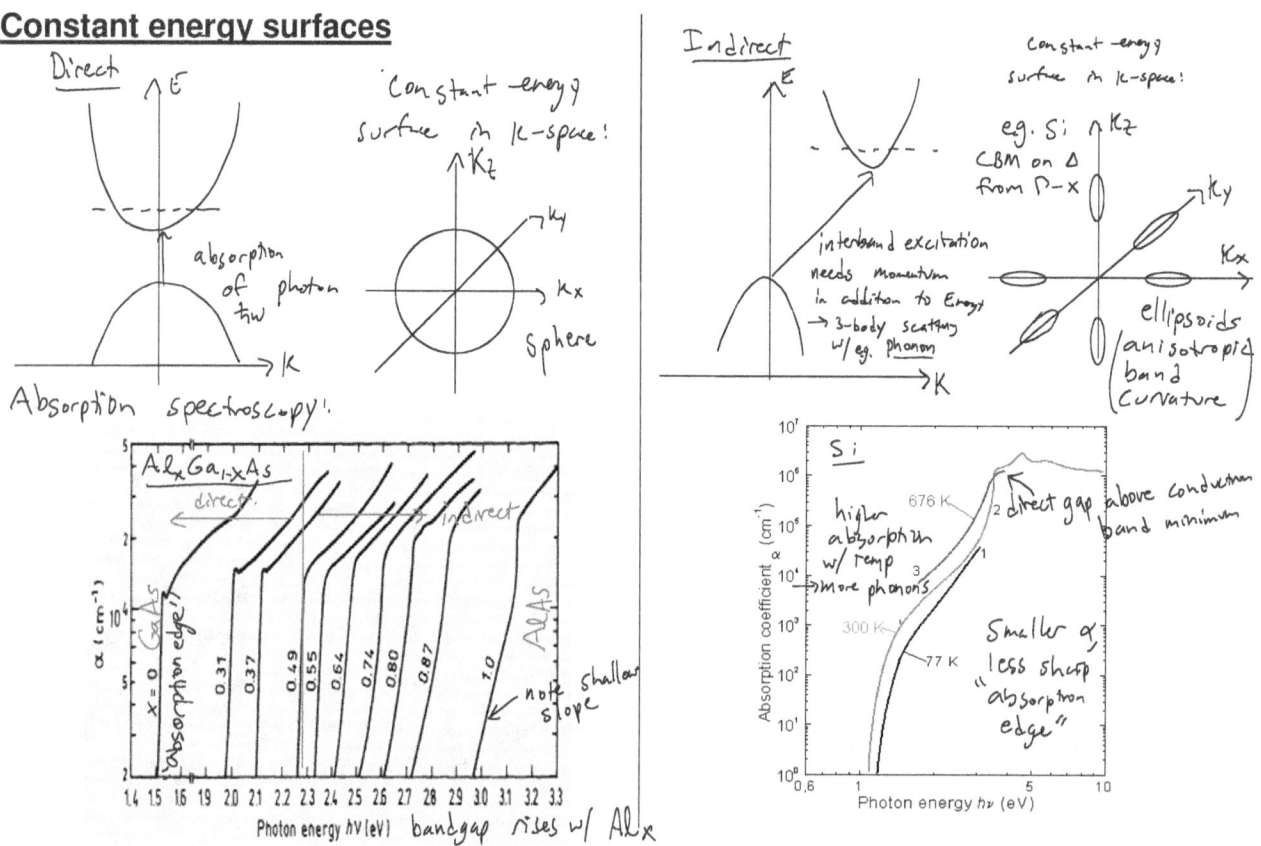

Electrons in bands: Dynamics

Velocity:
$v_g = \frac{d\omega}{dk} = \frac{1}{\hbar}\frac{dE}{dk}$. If $F = \frac{dp}{dt} = \hbar \frac{dk}{dt}$. Then $\Delta k = \frac{F}{\hbar}\Delta t$

But for periodic $E(k)$, this implies oscillatory motion: AC current from DC field!

We need $\Delta k > \frac{2\pi}{a} \to \frac{e\mathcal{E}}{\hbar}\tau > \frac{2\pi}{a} \to \mathcal{E} > \frac{h}{e}\frac{1}{a\tau}$ too big for small τ!

This "Bloch oscillation" only possible for large lattice const: periodic "superlattice".

acceleration:
$a = \frac{dv_g}{dt} = \frac{d}{dt}\left(\frac{d\omega}{dk}\right) = \frac{1}{\hbar}\frac{d}{dt}\frac{dE}{dk} = \frac{1}{\hbar}\frac{d}{dk}\frac{dE}{dt} = \frac{1}{\hbar}\frac{d}{dk}\left(\frac{dE}{dk}\frac{dk}{dt}\right) = \frac{1}{\hbar^2}\frac{d^2E}{dk^2}\frac{dp}{dt} = \frac{1}{\hbar^2}\frac{d^2E}{dk^2}F$

So, $F = \frac{\hbar^2}{\frac{d^2E}{dk^2}} a = m^* a$ ($m^* \equiv$ "effective mass" $\to$ NOT inertial/gravitational, but dynamic)

Check: for free electron, $E(k) = \frac{\hbar^2 k^2}{2m}$, $\frac{d^2E}{dk^2} = \frac{\hbar^2}{m}$ so $m^* = \frac{\hbar^2}{\frac{d^2E}{dk^2}} = m$ ✓

$\Rightarrow$ replace $m \to m^*$ in Sommerfeld/Drude relations

But be careful! Anisotropy means m^* is really a tensor!

Valence band excitations

Note that by this definition, valence band electrons have a negative mass and their mobility $\mu = \frac{e\tau}{m^*}$ is of opposite sign as conduction electrons! Does this mean current flows the "wrong" way? NO!

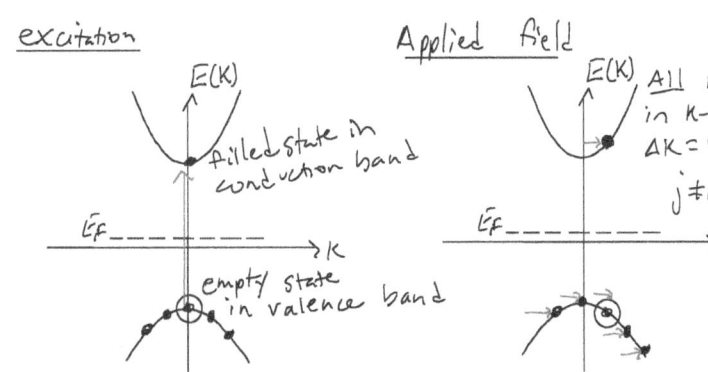

excitation / Applied field

Current density $j = nev_g$
Conduction band, $v_g^{cond.} \propto \frac{dE}{dk} > 0$
So $j_{cond.} > 0$

In valence band,
$v_g^{empty} \propto \frac{dE}{dk} < 0$, therefore
$j_{val.} \propto \sum_{all} v_g^{\to 0} - v_g^{empty} > 0$

Conceptual simplification:
"Hole": a positively-charged particle with positive effective mass, rather than the absence of a negatively-charged electron with a negative m^*. $\to$ simplifies calculation too!

Effective mass via perturbation theory: k dot p

Standard second-order perturbation theory gives

$$E_n(K) \cong E_n^0 + \underbrace{\langle \psi_n^0 | \mathcal{H}_p | \psi_n^0 \rangle}_{(1^{st})} + \underbrace{\sum_m \frac{|\langle \psi_n^0 | \mathcal{H}_p | \psi_m^0 \rangle|^2}{E_n - E_m}}_{(2^{nd})}$$

For bandstructure, our perturbation $\mathcal{H}_p = \frac{\hbar^2 k^2}{2m} + \frac{\hbar}{m} \vec{k} \cdot \vec{p}$

Two band model:

$$E_c(K) = E_c + \frac{\hbar^2 k^2}{2m} \underbrace{\langle \psi_c^0 | \psi_c^0 \rangle}_{1} + \frac{\hbar}{m} \underbrace{\langle \psi_c^0 | \vec{k} \cdot \vec{p} | \psi_c^0 \rangle}_{\substack{\to 0 \text{ (all odd powers} \\ \text{of } k \text{ give no contribution to } E(k) \\ \text{around extrema)}}}$$

$$+ \underbrace{\frac{\hbar^2 k^2}{2m} |\langle \psi_c^0 | \psi_v^0 \rangle|^2}_{E_c - E_v}^{\to 0} + \frac{\hbar^2}{m^2} \frac{|\langle \psi_c^0 | \vec{k} \cdot \vec{p} | \psi_v^0 \rangle|^2}{E_c - E_v}$$

$$= E_c + \frac{\hbar^2 k^2}{2m} + \frac{\hbar^2 k^2}{m^2} \frac{|\langle \psi_c^0 | p | \psi_v^0 \rangle|^2}{E_g} = E_c + \frac{\hbar^2 k^2}{2} \left(\frac{1}{m} + \frac{2\langle P \rangle^2}{m^2 E_g} \right) = E_c + \frac{\hbar^2 k^2}{2m^*}$$

where $\frac{1}{m^*} = \frac{1}{m} + \frac{2\langle P \rangle^2}{m^2 E_g}$ determines effective mass

Note that the denominator would be $E_v - E_c = -E_g$ for $E_v(K)$!

Approximate calculation of $\langle P \rangle^2$

In 1-D, we saw that the zone-edge wavefunctions above + below a bandgap were merely shifted in phase.

So, using $\psi_c^0 \propto \cos \frac{2\pi x}{a}$ and $\psi_v^0 \propto \sin \frac{2\pi x}{a}$,

$$\langle P \rangle = \langle \psi_c^0 | \frac{\hbar}{i} \frac{d}{dx} | \psi_v^0 \rangle = \frac{\hbar}{i} \cdot \frac{2\pi}{a} \underbrace{\langle \psi_c^0 | \psi_c^0 \rangle}_{\sim 1}$$

This gives $\langle P \rangle^2 \sim \frac{\hbar^2}{a^2}$ so that $\frac{1}{m^*} = \frac{1}{m} \left(1 + \frac{2\hbar^2}{m a^2 E_g} \right)$

If $\frac{2\hbar^2}{m a^2} \gg E_g$, then we can approximate

$$\frac{1}{m^*} \sim \frac{2\hbar^2}{m^2 a^2 E_g} \implies \frac{m^*}{m} = \frac{m a^2 E_g}{2 \hbar^2} \sim \frac{E_g}{E_p} \leftarrow \text{"Kane energy"} \quad \text{(approximate)}$$

Comparison to narrow-gap insulators

For common semiconductors (narrow-gap insulators), a typical lattice const $a \sim 5\text{-}6\,\text{Å}$. We then have

$$E_p \simeq \frac{2\hbar^2}{ma^2} \sim \frac{2(4\times 10^{-16}\,\text{eV·s})^2}{5\times 10^5\,\text{eV}/c^2 \cdot (5\times 10^{-8}\,\text{cm})^2} = \frac{2\cdot 16\times 10^{-30}\,\text{eV}^2 s^2 \cdot 9\times 10^{20}\,\text{cm}^2/s^2}{5\times 10^5\,\text{eV} \cdot 25\times 10^{-16}\,\text{cm}^2} \approx 20\,\text{eV}$$

Therefore,

$$\frac{m^*}{m} \sim \frac{E_g}{20\,\text{eV}}$$

This two-band perturbation theory result fits remarkably well to experiment!

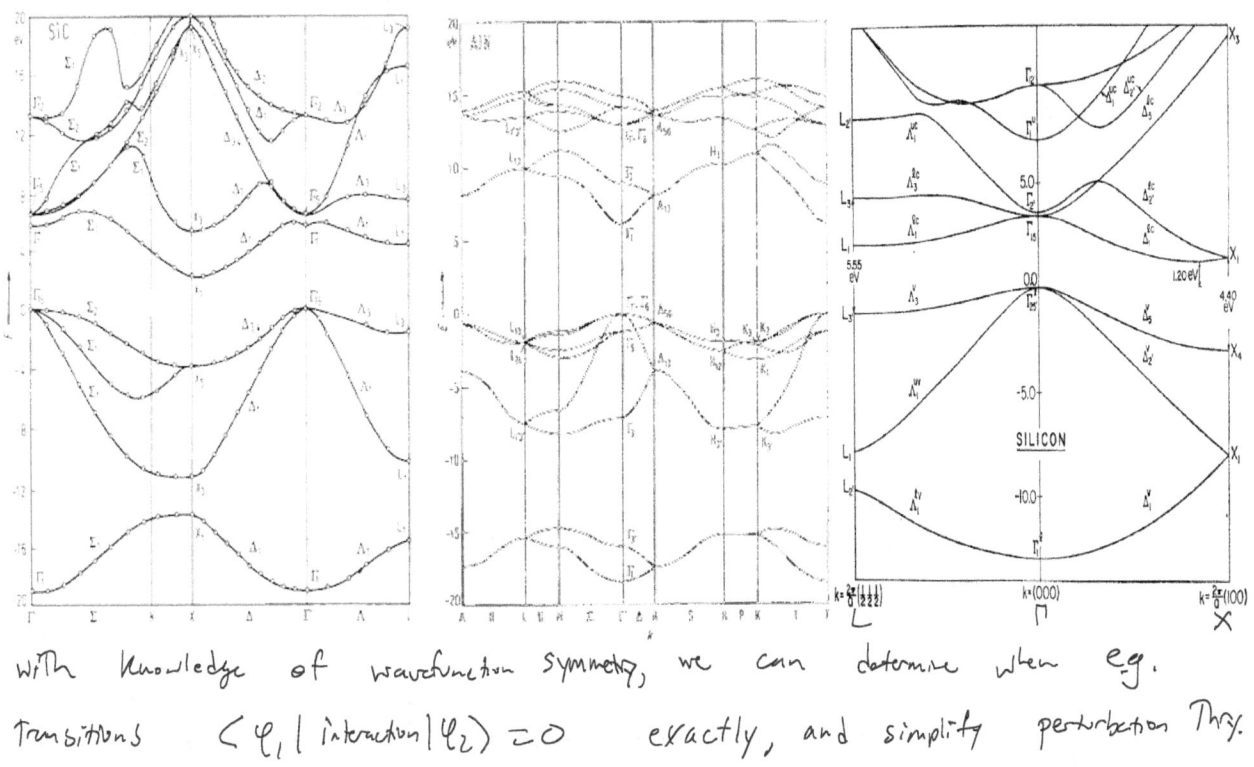

From eigenvalues to eigenfunctions: Or, "What are all those labels?!"

With knowledge of wavefunction symmetry, we can determine when e.g. transitions $\langle \psi_1 | \text{interaction} | \psi_2 \rangle = 0$ exactly, and simplify perturbation Thry.

71

Bare essentials of group theory

Symmetry operations which leave the Hamiltonian invariant must also leave the wavefunction invariant, or result in generation of a degenerate state.

$$\mathcal{H}\psi = E\psi \implies T\mathcal{H}\psi = TE\psi \implies \mathcal{H}(T\psi) = E(T\psi)$$

eg s-states (invariant) / p-states (3× degenerate) in coulomb potential (spherical sym)

So analysis of discrete crystal symmetry tells us about wavefunction symmetry. This analysis makes use of "group theory"

group: set of elements $\{a, b, ...\}$ including identity E and all inverses, where operation ab between any two elements is defined, and is associative and **closed** (i.e. the product yields a group element)

Example of point group: T_d

tetrahedron
e.g. methane or 4-fold coordinated diamond/zincblende

Elements:

- (×1) E identity
- (×3) C_2 2-fold (180°) rotation about $x, y,$ or z
- (×4) C_3 3-fold CW (120°) rotation about any "bond"
- (×4) C_3^{-1} 3-fold CCW (120°) rotation about any "bond"
- (×6) σ reflection about any plane intersecting 2 "atoms"
- (×3) S_4 4-fold CW (90°) rotation about $x, y,$ or z, and then reflection on a plane perpendicular to that axis
- (×3) S_4^{-1} 4-fold CCW (90°) rotation about $x, y,$ or z, and then reflection on a plane perpendicular to that axis

We can quantify these group elements using their **matrix** representations!

Representation

- Matrices can be used to represent the elements of a group. One way to generate these matrices is to define basis functions that transform into superpositions of themselves under group symmetry operations. The matrices which contain the transformed coefficients are a **representation**. Matrices generated in this way will have dimension equal to the number of group elements.

Example: CCW 4-fold rotation about x-axis

$$\begin{bmatrix} 1 & 0 & 0 \\ 0 & 0 & 1 \\ 0 & -1 & 0 \end{bmatrix} \begin{bmatrix} x \\ y \\ z \end{bmatrix} = \begin{bmatrix} x \\ z \\ -y \end{bmatrix}$$

another: CW 3-fold rotation about the $\hat{x}+\hat{y}+\hat{z}$ axis

$$\begin{bmatrix} 0 & 0 & 1 \\ 1 & 0 & 0 \\ 0 & 1 & 0 \end{bmatrix} \begin{bmatrix} x \\ y \\ z \end{bmatrix} = \begin{bmatrix} z \\ x \\ y \end{bmatrix}$$

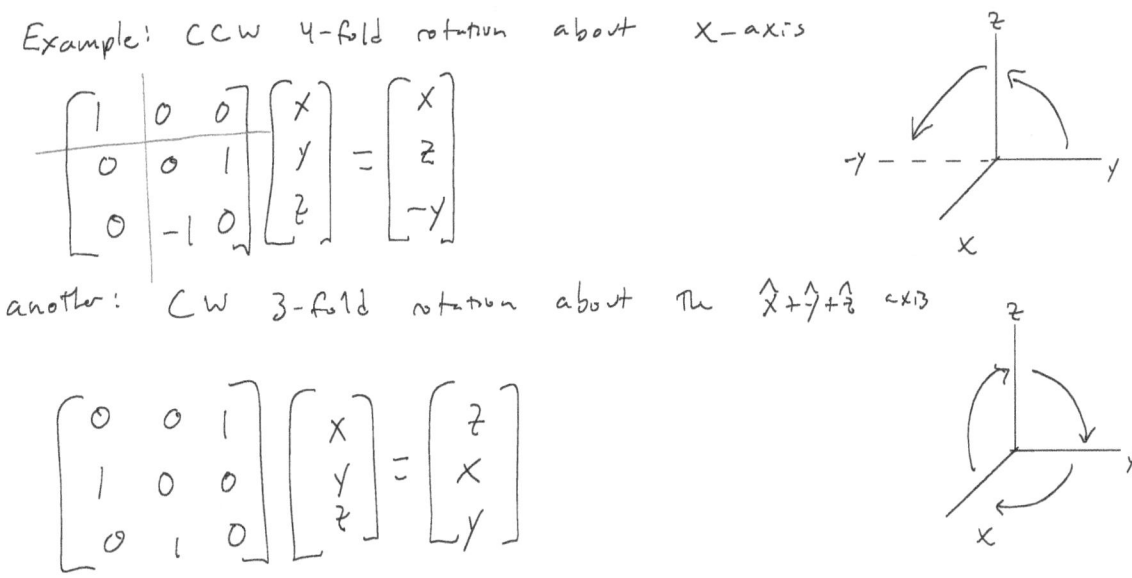

- If a unitary transformation of the matrices results in block-diagonal form, then the representation is **reducible** into smaller dimension. Although the choice of **irreducible** representation is not unique by a unitary transformation, the traces (sum of diagonal elements) are preserved. These traces are called **characters**.

Character table

- All elements in the same **class** have the same character. Classes consist of the set of symmetry elements that are transformed into one another by unitary transformations from the other group elements. *The number of irreducible representations is equal to the number of classes.* Therefore, **a character table** can be formed by a square array with classes horizontally and representations vertically.
- The vertical sum over representations of the squared norms of characters in each class is equal to the number of group elements divided by the number of class elements. The sum over representations of the products of characters from different classes is zero. (orthogonality #1)
- The horizontal sum over classes of the squared norm of characters times number of class elements is equal to the number of group elements. The sum over classes of products of different representations times number of class elements is zero. (orthogonality #2)

		{E}	{3C₂}	{6S₄}	{6σ}	{8C₃}	= 24 elements
(A₁)	Γ₁	1	1	1	1	1	$1\cdot 1^2+3\cdot 1^2+6\cdot 1^2+6\cdot 1^2+8\cdot 1^2=24$
(A₂)	Γ₂	1	1	-1	-1	1	
(E)	Γ₃	2	2	0	0	-1	$1\cdot 2^2+3\cdot 2^2+8\cdot(-1)^2=24$
(T₁)	Γ₅	3	-1	1	-1	0	
(T₂)	Γ₄	3	-1	-1	1	0	$1\cdot 3^2+3\cdot 1^2+6\cdot 1^2+6\cdot 1^2=24$

Column sums of squared characters:
- $=1^2+1^2+2^2+3^2+3^2=24$
- $=1^2+1^2+2^2+3^2+3^2=8=24/3$
- $=1^2+1^2+1^2+1^2=4=24/6$
- $=1^2+1^2+1^2=3=24/8$

- The traces of the identity element class are equal to the dimension of the representation, so #1 above can be used to complete an entire column for the identity {E}.
- There is always a trivial identity representation with all characters unity.

Wavefunction symmetry at k-points along IBZ path

- Each reciprocal lattice point will have its own transformation properties under symmetry operations. Determine the character tables for their reducible representations by finding the number of k-points unchanged by each operation

- Reducible representations can be decomposed into direct (Kronecker) products and direct sums (block diagonal concatenation) of irreducible representations using the character table, since the product of traces is the trace of the direct product and the sum of the traces is the trace of the direct sum.

Example: $\vec{k} = \frac{2\pi}{a}(1,1,1)$ has 8 equivalent points

{E}	{3C$_2$}	{6S$_4$}	{6σ}	{8C$_3$}
8	0	0	4	2

decomposed into $2\Gamma_1 \oplus 2\Gamma_4 = \begin{matrix} 2 \times [1\ 1\ 1\ 1\ 1] \\ + 2 \times [3\ -1\ -1\ 1\ 0] \end{matrix}$

Likewise, $\vec{k} = \frac{2\pi}{a}(2,0,0)$ has 6 equivalent points

{E}	{3C$_2$}	{6S$_4$}	{6σ}	{8C$_3$}
6	2	0	2	0

decomposed into $\Gamma_1 \oplus \Gamma_3 \oplus \Gamma_4 = \begin{matrix} [1\ 1\ 1\ 1\ 1] \\ + [2\ 2\ 0\ 0\ -1] \\ + [3\ -1\ -1\ 1\ 0] \end{matrix}$

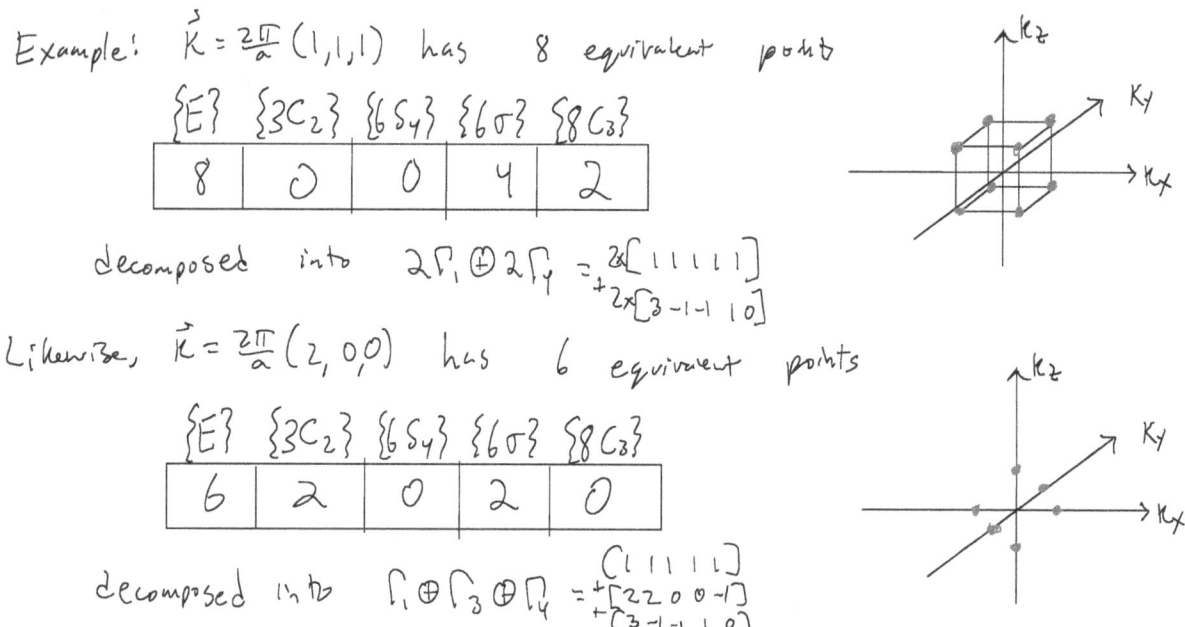

Band symmetry labels: zincblende

We can now label free electron states at Γ with corresponding representations. This indicates degeneracy and provides a means to determine how they split due to atomic potential.

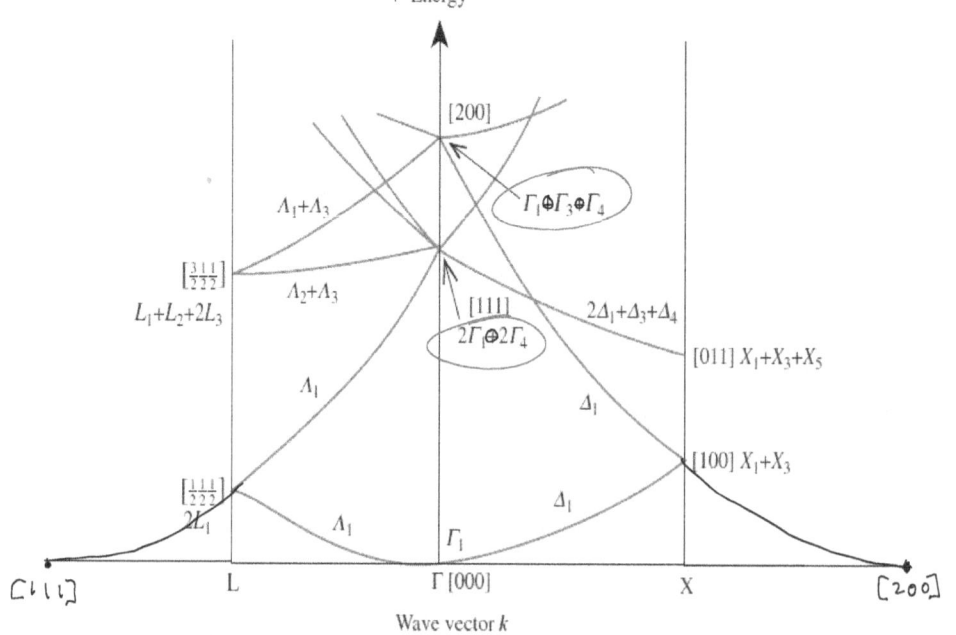

Decomposition of representation: matrix elements

- **matrix element theorem**: Matrix element of operator p between Ψ_1 and Ψ_2 can only be nonzero when the direct product of two representations can be decomposed into a direct sum containing the representation of the third.

Example: (trivial) point group of 1D functions

Character table: $\{E\}$ $\{\sigma\}$ (reflection)

	$\{E\}$	$\{\sigma\}$
"even" S	1	1
"odd" A	1	-1

$1 \cdot 1^2 + 1 \cdot 1^2 = 2$

$1^2 + 1^2 = 2$

operator S:
$S \otimes S = S$
$S \otimes A = A$

operator A:
$A \otimes S = A$
$A \otimes A = S$

So matrix elements of operators corresponding to representation "S" between wavefunctions of __different__ representation are zero, and matrix elements of operators of representation "A" between wavefunctions of the __same__ representation are zero....

But you already know this!: integral of an __odd__ integrand over symmetric bounds $\langle \varphi_1 | \mathcal{O} | \varphi_2 \rangle = 0$!!

Basis functions

Notice the functions usually appended to the right of the character table. These are a particular choice of basis functions that transform into each other under the group operations.

T_d		E	$8C_3$	$3C_2$	$6\sigma_d$	$6S_4$		
Γ_1	A_1	1	1	1	1	1		$x^2+y^2+z^2 (=r^2)$
Γ_2	A_2	1	1	1	-1	-1		
Γ_3	E	2	-1	2	0	0		$(x^2-y^2, 3z^2-r^2)$
Γ_5	T_1	3	0	-1	-1	1	(I_x, I_y, I_z)	
Γ_4	T_2	3	0	-1	1	-1	(x, y, z)	(xy, yz, zx)

They can be compared to the symmetry of operator terms in the Hamiltonian. Here, we see that any scalar has Γ_1 symmetry, and any polar vector (x,y,z) or axial vector (xy, yz, zx) has Γ_4 symmetry.

Example: Electric dipole transitions

The electron-radiation interaction Hamiltonian is $\frac{e}{m}\vec{A}\cdot\vec{p} \to e\vec{r}\cdot\vec{\mathcal{E}}$ (in dipole approximation). $\vec{p}$ transforms like a polar vector, so has Γ_4 symmetry in T_d. Calculate products of characters:

$\Gamma_4 \otimes \Gamma_1 =$	3	-1	-1	1	0	$= \Gamma_4$ ✓
$\Gamma_2 =$	3	-1	1	-1	0	$= \Gamma_5$ ← no Γ_4! ✓
$\Gamma_3 =$	6	-2	0	0	0	$= \Gamma_4 \oplus \Gamma_5$ ✓
$\Gamma_4 =$	9	1	1	1	0	$= \Gamma_4 \oplus \Gamma_5 \oplus \Gamma_3 \oplus \Gamma_1$ ✓
$\Gamma_5 =$	9	1	-1	-1	0	$= \Gamma_4 \oplus \Gamma_5 \oplus \Gamma_3 \oplus \Gamma_2$ ✓

The valence band of group IV and III-V semiconductors is composed mostly from the 3-fold degenerate "p" atomic orbitals → They are basis functions of the irreducible representation Γ_4, with dimension 3! So, electric dipole transitions are allowed from this Γ_4 valence band state to **all** conduction band states, except to those of Γ_2 **by symmetry**!

Example: longitudinal effective mass in Si

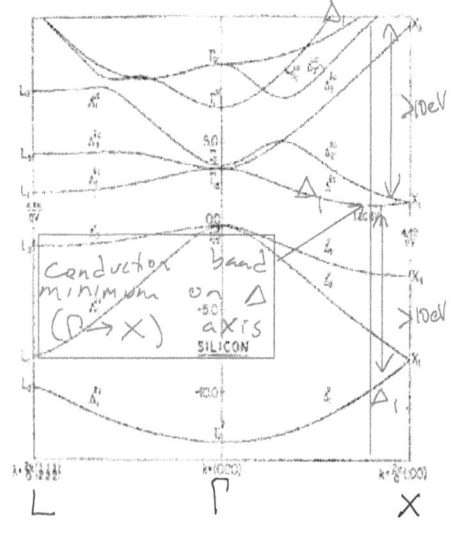

L   Γ   X

$$\frac{1}{m_\ell^*} = \frac{1}{m_0} + \frac{2}{m_0^2} \sum_n \frac{|\langle \psi_c | p_z | \psi_n \rangle|^2}{\mathcal{E}_c - \mathcal{E}_n}$$

Bandstructure shows ψ_c has Δ_1 symmetry. From character table of Δ, we see p_z (component of polar vector along valley axis) transforms like Δ_1 as well.

representation	basis functions	E	C_4^2	$2C_4$	$2iC_4^2$	$2iC_2'$
Δ_1	$z, 2x^2-y^2-z^2$	1	1	1	1	1
Δ_2	y^2-z^2	1	1	-1	1	-1
Δ_2'	yz	1	1	-1	-1	1
Δ_1'	$yz(y^2-z^2)$	1	1	1	-1	-1
Δ_5	xy, xz	2	-2	0	0	0

Since $\Delta_1 \otimes \Delta_1 = \Delta_1$, all matrix elements are identically zero, except those that couple the Δ_1 lowest conduction band with another state of Δ_1 symmetry. But those are >10 eV away, so they make negligible contribution → $m_\ell^* = m_0$!

"Tight binding" / "LCAO"

Construct wavefunctions from the isolated atomic orbitals

Example: 1 atomic s-orbital on 1-D lattice

$$\Psi(r) = \sum_m C_m |\phi_s(r-R_m)\rangle \qquad |C_m|^2 \text{ is independent of } m$$

$|\phi_s(r-R_m)\rangle$ are "Löwdin" orbitals chosen to be orthogonal on different sites.

Must be consistent w/ Bloch condition: $\Psi(r+a) = e^{ika}\Psi(r)$

$$\Psi(r) = \sum_m \frac{e^{ikR_m}}{\sqrt{N}} |\phi_s(r-R_m)\rangle \rightarrow \Psi(r+a) = \sum_m \frac{e^{ikR_m}}{\sqrt{N}} |\phi_s(r+a-R_m)\rangle$$

Now, variable transformation $R_{m'} = R_m - a$ = $\sum_{m'} \frac{e^{ik(R_{m'}+a)}}{\sqrt{N}} |\phi_s(r-R_{m'})\rangle = e^{ika}\Psi(r)$

So it is a valid basis ....

Substitution into Schrodinger Equation

To calculate energy spectrum (bands), substitute into Schrodinger Eq.

$$\sum_m \frac{e^{ikR_m}}{\sqrt{N}} \mathcal{H} |\phi_s(r-R_m)\rangle = E \sum_m \frac{e^{ikR_m}}{\sqrt{N}} |\phi_s(r-R_m)\rangle$$

where $\mathcal{H} = \mathcal{H}_0 + \mathcal{H}_{int}$. The first term is just the sum of atomic potentials. The second is the interatomic interaction to be treated perturbatively.

Energy of one unit cell is calculated by projection onto $\langle\phi_s(r)|$

$$\frac{e^{ika}}{\sqrt{N}} \underbrace{\langle\phi_s(r)|\mathcal{H}_{int}|\phi_s(r-a)\rangle}_{-V_{ss\sigma}} + \frac{1}{\sqrt{N}} \underbrace{\langle\phi_s(r)|\mathcal{H}_0|\phi_s(r)\rangle}_{E_s} + \frac{e^{ika}}{\sqrt{N}} \underbrace{\langle\phi_s(r)|\mathcal{H}_{int}|\phi_s(r+a)\rangle}_{-V_{ss\sigma}} = \frac{E}{\sqrt{N}} \underbrace{\langle\phi_s(r)|\phi_s(r)\rangle}_{1}$$

+ negligible terms from non-nearest-neighbors and diagonal term of $\mathcal{H}_{int}$

+ vanishing terms (orthogonality)

This gives our dispersion directly:

$$E(k) = -V_{ss\sigma}(e^{ika} + e^{-ika}) + E_s$$
$$= E_s - 2V_{ss\sigma}\cos ka$$

Comparison to "Finite Differences" Hamiltonian

Use symmetric definition of derivative
$$\frac{d}{dx}\psi(x) = \lim_{\Delta x \to 0} \frac{\psi(x+\frac{\Delta x}{2}) - \psi(x-\frac{\Delta x}{2})}{\Delta x}$$

Then,
$$\frac{d^2}{dx^2}\psi(x) = \lim_{\Delta x \to 0} \frac{\frac{\psi(x+\Delta x)-\psi(x)}{\Delta x} - \frac{\psi(x)-\psi(x-\Delta x)}{\Delta x}}{\Delta x} = \lim_{\Delta x \to 0} \frac{\psi(x-\Delta x) - 2\psi(x) + \psi(x+\Delta x)}{\Delta x^2}$$

Now, instead of passing to limit $\Delta x \to 0$ to recover continuous differential operator, use "Finite differences": Approximate derivatives by using nonzero value of Δx i.e. <u>discretize</u> continuous variable x : N segments from $x=0$ to $x=L$:

Then,
$$\frac{d^2\psi(x_i)}{dx^2} \sim \frac{\psi(x_{i-1}) - 2\psi(x_i) + \psi(x_{i+1})}{\Delta x^2} \equiv \frac{\psi_{i-1} - 2\psi_i + \psi_{i+1}}{\Delta x^2} \equiv \hat{D}\vec{\psi}$$

$\hat{D}$ is a <u>tridiagonal</u> matrix, i.e. only nearest-neighbor coupling, just like our LCAO Hamiltonian, with $\frac{\hbar^2}{2m\Delta x^2} \to \frac{E_s}{2}$ (diagonal) , $V_{ss\sigma}$ (off-diagonal)

The fact that, in general, $\frac{E_s}{2} \neq V_{ss\sigma}$ means that $m^* \neq m_0$.

Comparison with the nearly-free electron bands

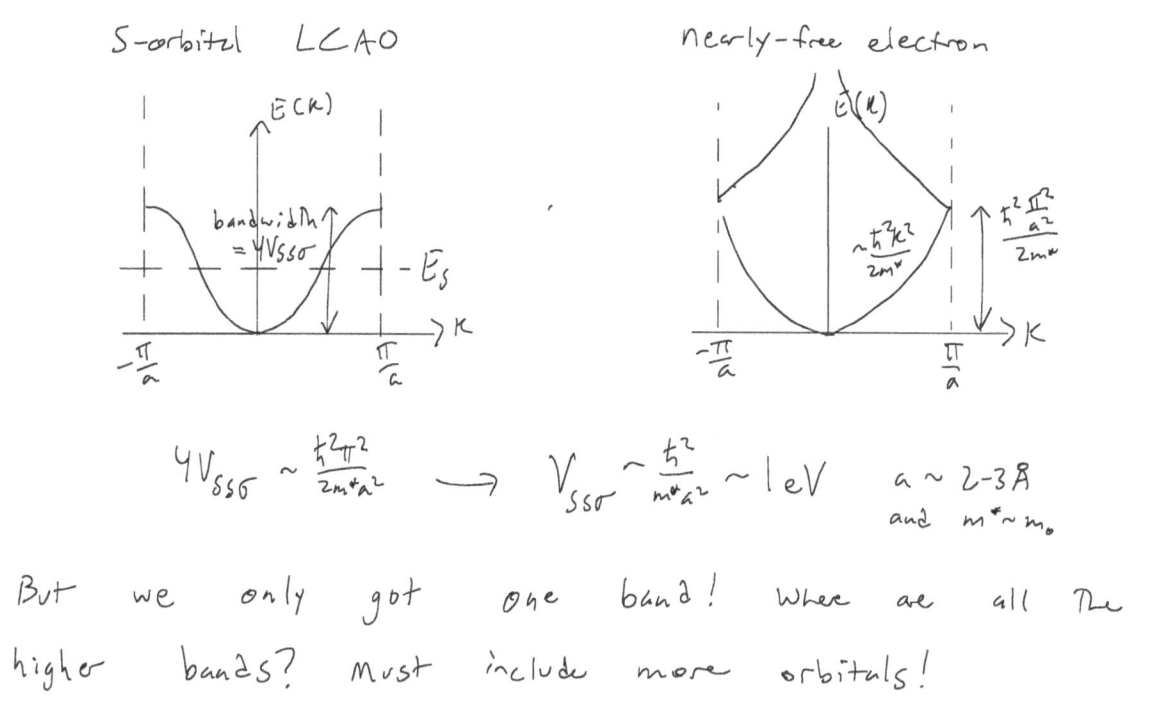

$$4V_{ss\sigma} \sim \frac{\hbar^2 \pi^2}{2m^* a^2} \longrightarrow V_{ss\sigma} \sim \frac{\hbar^2}{m^* a^2} \sim 1\,eV \quad a \sim 2-3\,Å \text{ and } m^* \sim m_0$$

But we only got one band! Where are all the higher bands? Must include more orbitals!

2-band LCAO: "sp" model

Now expand in two orbitals:

$$\psi(r) = \sum_m \frac{e^{ikR_m}}{\sqrt{N}} \left[C_s |\phi_s(r-R_m)\rangle + C_p |\phi_p(r-R_m)\rangle \right]$$

Insert into Schrödinger Eqn and project onto basis states at origin

$$\langle \phi_s(r)| \times \begin{bmatrix} e^{ika}\mathcal{H}|\phi_s(r-a)\rangle C_s + e^{ika}\mathcal{H}|\phi_p(r-a)\rangle C_p \\ + e^{-ika}\mathcal{H}|\phi_s(r+a)\rangle C_s + e^{-ika}\mathcal{H}|\phi_p(r+a)\rangle C_p = E\left(|\phi_s(r)\rangle^1 C_s + |\phi_p(r)\rangle^0 C_p\right) \\ + \mathcal{H}|\phi_s(r)\rangle C_s + \mathcal{H}|\phi_p(r)\rangle^0 C_p \end{bmatrix}$$

$$\left[E_s - 2V_{ss\sigma}\cos ka\right]C_s + \left[e^{ika}\underbrace{\langle\phi_s(r)|\mathcal{H}|\phi_p(r-a)\rangle}_{V_{sp\sigma}} + e^{-ika}\underbrace{\langle\phi_s(r)|\mathcal{H}|\phi_p(r+a)\rangle}_{-V_{sp\sigma}}\right]C_p = EC_s$$

$$\left[E_s - 2V_{ss\sigma}\cos ka\right]C_s + \left[2iV_{sp\sigma}\sin ka\right]C_p = EC_s \qquad \text{①}$$

p-state projection

$$\langle \phi_p(r)| \times \begin{bmatrix} e^{ika}\mathcal{H}|\phi_s(r-a)\rangle C_s + e^{ika}\mathcal{H}|\phi_p(r-a)\rangle C_p \\ + e^{-ika}\mathcal{H}|\phi_s(r+a)\rangle C_s + e^{-ika}\mathcal{H}|\phi_p(r+a)\rangle C_p = E\left(|\phi_s(r)\rangle^0 C_s + |\phi_p(r)\rangle^1 C_p\right) \\ + \mathcal{H}|\phi_s(r)\rangle^0 C_s + \mathcal{H}|\phi_p(r)\rangle C_p \end{bmatrix}$$

$$E_p C_p + \left(e^{ika}\underbrace{\langle\phi_p(r)|\mathcal{H}|\phi_p(r-a)\rangle}_{V_{pp\sigma}} + e^{-ika}\underbrace{\langle\phi_p(r)|\mathcal{H}|\phi_p(r+a)\rangle}_{V_{pp\sigma}}\right)C_p$$
$$+ \left(e^{ika}\underbrace{\langle\phi_p(r)|\mathcal{H}|\phi_s(r-a)\rangle}_{=-V_{sp\sigma}} + e^{-ika}\underbrace{\langle\phi_p(r)|\mathcal{H}|\phi_s(r+a)\rangle}_{=V_{sp\sigma}}\right)C_s = EC_p$$

$$\left[E_p + 2V_{pp\sigma}\cos ka\right]C_p + \left[-2iV_{sp\sigma}\sin ka\right]C_s = EC_p \qquad \text{②}$$

2x2 matrix eigenvalue problem

Equations ① and ② are equivalent to:

$$\begin{bmatrix} E_s - 2V_{ss\sigma}\cos ka & 2iV_{sp\sigma}\sin ka \\ -2iV_{sp\sigma}\sin ka & E_p + V_{pp\sigma}\cos ka \end{bmatrix} \begin{bmatrix} C_s \\ C_p \end{bmatrix} = E \begin{bmatrix} C_s \\ C_p \end{bmatrix}$$

Note that our 2x2 Hamiltonian matrix is <u>hermitian</u> as required for real E and <u>diagonal</u> for $k=0$ and $\pm \frac{\pi}{a}$. This means eigenvectors are $\begin{bmatrix}1\\0\end{bmatrix}$ and $\begin{bmatrix}0\\1\end{bmatrix}$, i.e. <u>purely</u> s- or p-orbitals! At k-points interior to the Brillouin zone, we have a mixture for $V_{sp\sigma} \neq 0$.

Numerical results (Two orbitals, two bands)

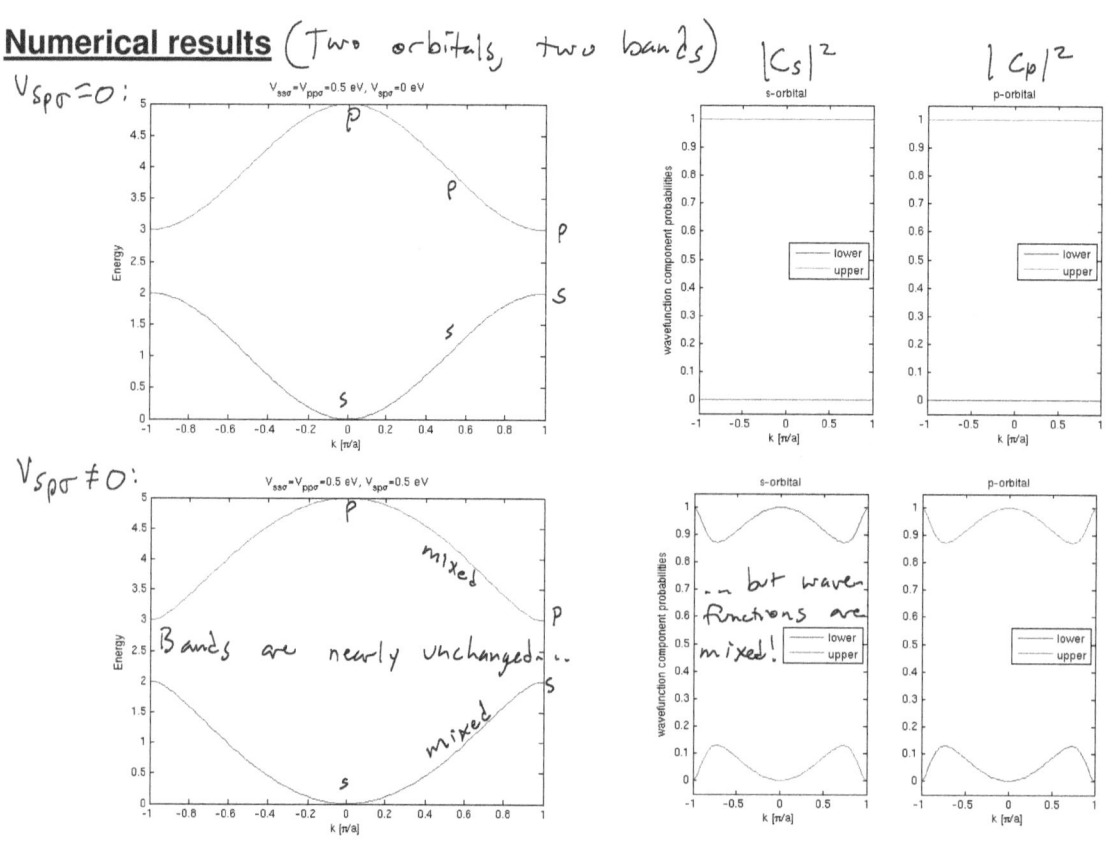

$V_{sp\sigma} = 0$:

$V_{sp\sigma} \neq 0$: Bands are nearly unchanged... mixed, mixed ... but wavefunctions are mixed!

Band mixing and band extrema at interior k-points

If bandwidth exceeds on-site energy, band crossings occur when off-diagonal $V_{sp\sigma}=0$. But, states are still purely s or p:

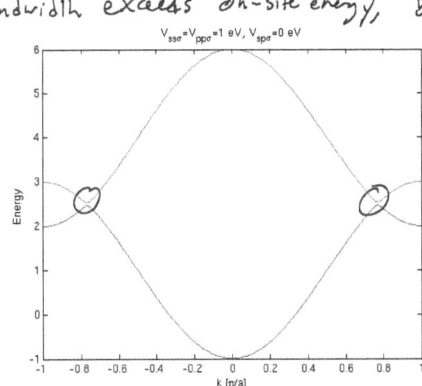
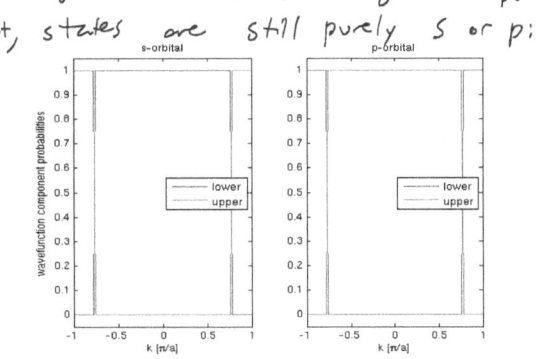

When $V_{sp\sigma} \neq 0$, avoided crossings give extrema at interior k-points and mix s and p

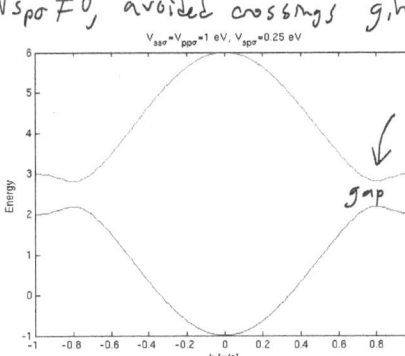
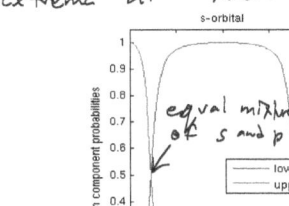
gap
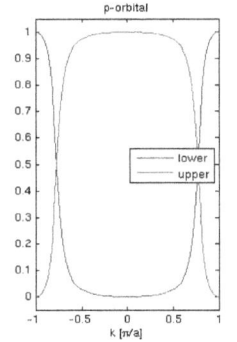
equal mixture of s and p

Tight binding bandstructure in 2D: graphene

Each group-IV carbon atom hybridizes two p and one s orbital to form 3 sp^2 bonds in x-y plane. This gives "Honeycomb" lattice: 2 superposed triangular Bravais lattices:

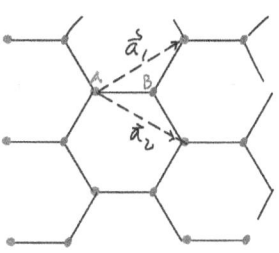

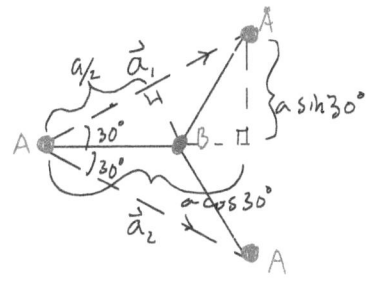

$|\vec{a}_1|, |\vec{a}_2| = a$

lattice vectors:

$\vec{a}_1 = \frac{\sqrt{3}}{2}a\hat{x} + \frac{a}{2}\hat{y}$

$\vec{a}_2 = \frac{\sqrt{3}}{2}a\hat{x} - \frac{a}{2}\hat{y}$

basis vectors:

$\vec{d}_1 = \vec{0}$

$\vec{d}_2 = \frac{a/2}{\cos 30°}\hat{x} = \frac{a}{\sqrt{3}}\hat{x}$

Nearest-neighbor vectors:

from A:

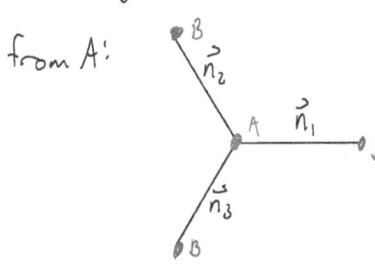

$\vec{n}_1 = \vec{d}_2$
$\vec{n}_2 = \vec{d}_2 - \vec{a}_2$
$\vec{n}_3 = \vec{d}_2 - \vec{a}_1$

from B:

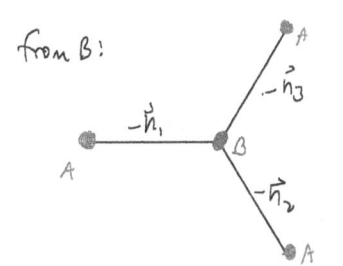

Tight binding wavefunction: Basis

The local triply-coordinated bonding leaves one unpaired electron in an out-of-plane p_z orbital which does not overlap with the in-plane p_x and p_y orbitals. Two atoms per unit cell (on sublattices A, B) form a "π bond":

Our wavefunction constructed from this two-orbital basis is

$$\Psi(\vec{r}) = \sum_m \frac{e^{i\vec{K}\cdot\vec{R}_m}}{\sqrt{N}} \left[C_A e^{i\vec{K}\cdot\vec{d}_1} \phi_{p_A}(\vec{r}-\vec{R}_m-\vec{d}_1) + C_B e^{i\vec{K}\cdot\vec{d}_2} \phi_{p_B}(\vec{r}-\vec{R}_m-\vec{d}_2) \right]$$

Schrodinger Eqn

Include **only** nearest neighbors, use orthogonality, neglect vanishing inner products

$$\langle \phi_{p_A} | \times \left[\frac{\mathcal{H}}{\sqrt{N}} \begin{Bmatrix} e^{i\vec{K}\cdot\vec{n}_1} C_B | \phi_{p_B}(\vec{r}-\vec{n}_1)\rangle \\ + e^{i\vec{K}\cdot\vec{n}_2} C_B | \phi_{p_B}(\vec{r}-\vec{n}_2)\rangle \\ + e^{i\vec{K}\cdot\vec{n}_3} C_B | \phi_{p_B}(\vec{r}-\vec{n}_3)\rangle \\ + C_A | \phi_{p_A}(\vec{r})\rangle \end{Bmatrix} = E\Psi \right] = E \frac{C_A}{\sqrt{N}} \underbrace{\langle \phi_{p_A} | \phi_{p_A}\rangle}_{1}$$

$\mathcal{H} = \mathcal{H}_0 + \mathcal{H}_{int}$.

Define $\langle \phi_{p_A}(\vec{r}) | \mathcal{H}_0 | \phi_{p_A}(\vec{r}) \rangle = E_p$ and $\langle \phi_{p_A}(\vec{r}) | \mathcal{H}_{int} | \phi_{p_B}(\vec{r}-\vec{n}) \rangle = -V_{pp\pi}$

Due to cylindrical symmetry about $\hat{z}$, all off-diagonal elements are equal!

Matrix eigenvalue problem

$E_p C_A - V_{pp\pi} C_B f(k) = E C_A$, where $f(k) = e^{i\vec{k}\cdot\vec{n}_1} + e^{i\vec{k}\cdot\vec{n}_2} + e^{i\vec{k}\cdot\vec{n}_3}$

Now project onto B-site orbital $\langle \phi_{pB}(\vec{r})|$, resulting in subscript $A \to B$ and $\vec{n} \to -\vec{n}$

$E_p C_B - V_{pp\pi} C_A f^*(k) = E C_B$

Equivalently,

$$\begin{array}{c} \langle p_A| \\ \langle p_B| \end{array} \begin{bmatrix} \overset{|p_A\rangle}{E_p} & \overset{|p_B\rangle}{-V_{pp\pi} f(k)} \\ -V_{pp\pi} f^*(k) & E_p \end{bmatrix} \begin{bmatrix} C_A \\ C_B \end{bmatrix} = E \begin{bmatrix} C_A \\ C_B \end{bmatrix}$$

Eigenvalues $(E_p - E)^2 - V_{pp\pi}^2 |f(k)|^2 = 0 \longrightarrow E = E_p \pm V_{pp\pi}|f(k)|$

Now just need to define $\vec{k}$ along IBZ perimeter in reciprocal space!

Reciprocal lattice

reciprocal lattice vectors must satisfy $\vec{a}_i \cdot \vec{b}_j = 2\pi \delta_{ij}$, so we have 4 eqns

$a_1^x b_1^x + a_1^y b_1^y = 2\pi$ $a_2^x b_1^x + a_2^y b_1^y = 0$

$a_1^x b_2^x + a_1^y b_2^y = 0$ $a_2^x b_2^x + a_2^y b_2^y = 2\pi$

With previously-defined real-space lattice vectors:

$\vec{a}_1 = \frac{\sqrt{3}}{2} a \hat{x} + \frac{a}{2} \hat{y}$

$\vec{a}_2 = \frac{\sqrt{3}}{2} a \hat{x} - \frac{a}{2} \hat{y}$

we derive

$\vec{b}_1 = (\frac{1}{\sqrt{3}}\hat{x} + \hat{y})\frac{2\pi}{a} = (\frac{1}{2}\hat{x} + \frac{\sqrt{3}}{2}\hat{y})\frac{4\pi}{\sqrt{3}a}$

$\vec{b}_2 = (\frac{1}{\sqrt{3}}\hat{x} - \hat{y})\frac{2\pi}{a} = (\frac{1}{2}\hat{x} - \frac{\sqrt{3}}{2}\hat{y})\frac{4\pi}{\sqrt{3}a}$

So the reciprocal lattice is rotated 30° from real-space lattice!

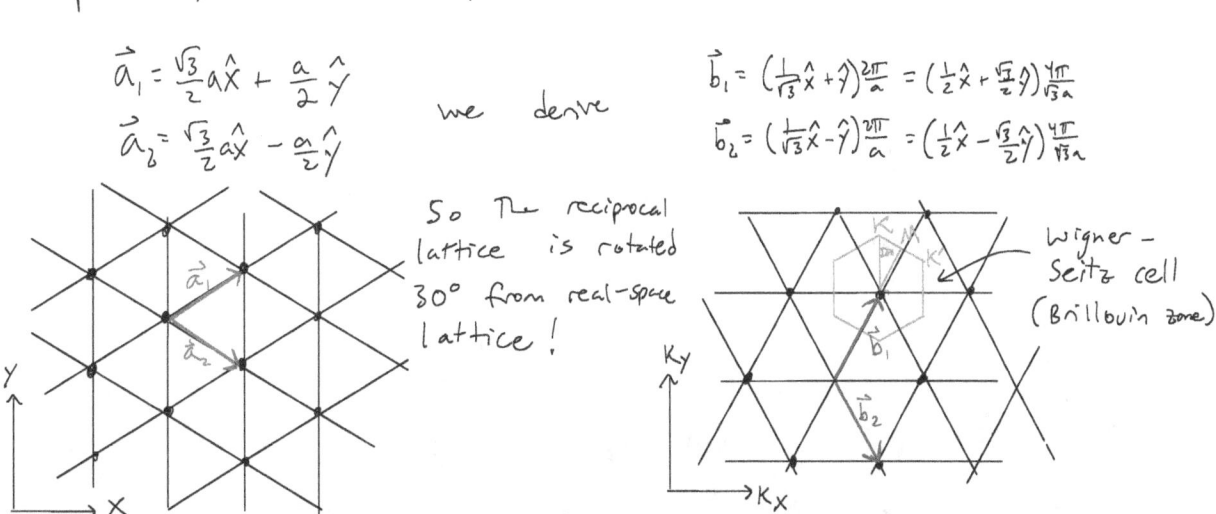

Wigner–Seitz cell (Brillouin zone)

Numerical results

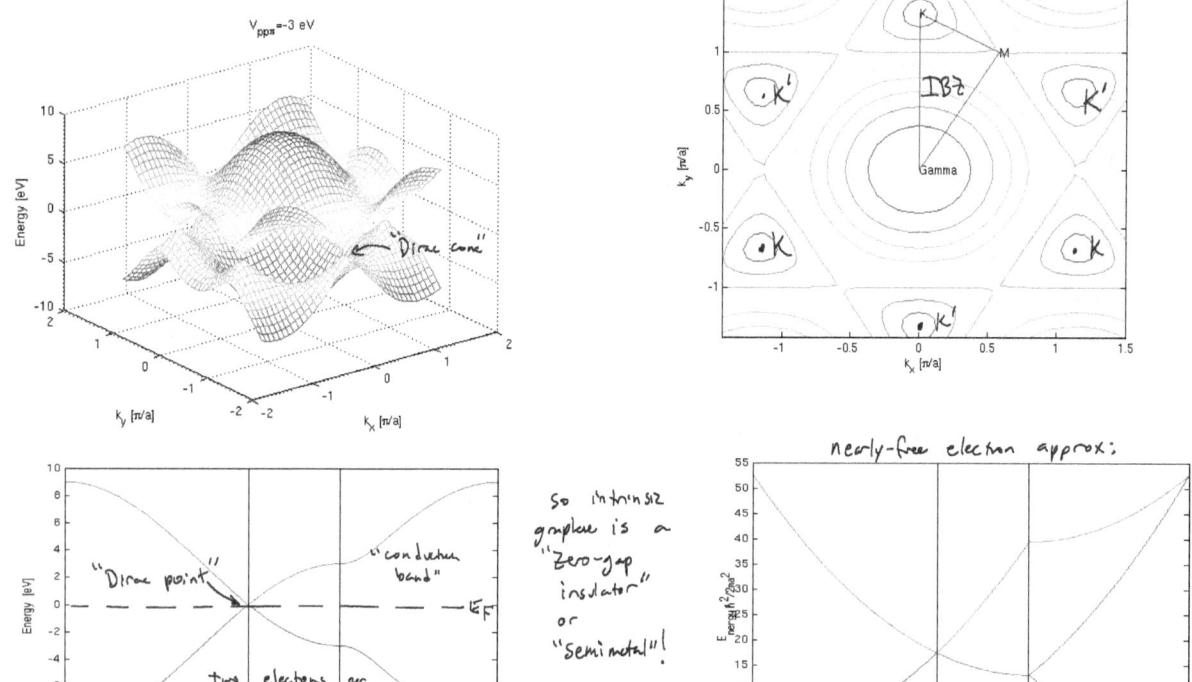

So intrinsic graphene is a "zero-gap insulator" or "semimetal"!

Near the Dirac point

$$f(\vec{K}+\vec{\Delta k}) = e^{i\left(\Delta k_x \hat{x} + \left(\frac{4\pi}{3a}+\Delta k_y\right)\hat{y}\right)} \cdot \begin{cases} \vec{n}_1 = \frac{a}{\sqrt{3}}\hat{x} \\ \vec{n}_2 = -\frac{a}{2\sqrt{3}}\hat{x} + \frac{a}{2}\hat{y} \\ \vec{n}_3 = -\frac{a}{2\sqrt{3}}\hat{x} - \frac{a}{2}\hat{y} \end{cases} \quad (+)$$

$$= e^{i\frac{\Delta k_x a}{\sqrt{3}}} + e^{i\left(\frac{2\pi}{3}+\frac{\Delta k_y a}{2}-\frac{\Delta k_x a}{2\sqrt{3}}\right)} + e^{i\left(\frac{2\pi}{3}-\frac{\Delta k_y a}{2}-\frac{\Delta k_x a}{2\sqrt{3}}\right)}$$

$$= e^{i\frac{\Delta k_x a}{\sqrt{3}}} + 2e^{-i\frac{\Delta k_x a}{2\sqrt{3}}} \cos\left(\frac{2\pi}{3}+\frac{\Delta k_y a}{2}\right)$$

$$= e^{i\frac{\Delta k_x a}{\sqrt{3}}} + 2e^{-i\frac{\Delta k_x a}{2\sqrt{3}}}\left(\cos\frac{2\pi}{3}\cos\frac{\Delta k_y a}{2} - \sin\frac{2\pi}{3}\sin\frac{\Delta k_y a}{2}\right)$$

$$\approx 1 + \frac{i\Delta k_x a}{\sqrt{3}} + 2\left(1 - \frac{i\Delta k_x a}{2\sqrt{3}}\right)\left(-\frac{1}{2}\cdot 1 - \frac{\sqrt{3}}{2}\frac{\Delta k_y a}{2}\right) \quad \text{(to first order)}$$

$$\approx \frac{i3\Delta k_x a}{2\sqrt{3}} - \frac{\sqrt{3}\Delta k_y a}{2} = \frac{\sqrt{3}a}{2}\left(i\Delta k_x - \Delta k_y\right)$$

$$\text{So } \mathcal{H} \approx \begin{bmatrix} E_p & V_{pp\pi}\frac{\sqrt{3}a}{2}(-i\Delta k_x + \Delta k_y) \\ V_{pp\pi}\frac{\sqrt{3}a}{2}(i\Delta k_x + \Delta k_y) & E_p \end{bmatrix} \quad \text{near Dirac point.}$$

84

Wavefunction components: pseudospin

On a path along circumference of a circle centered at Dirac point $\Delta k_y - i\Delta k_x = |\Delta k| e^{i\theta}$. Then,

$$\mathcal{H} \approx \begin{bmatrix} E_p & z e^{i\theta} \\ z e^{-i\theta} & E_p \end{bmatrix}$$

with eigenvalues $E_p \pm |z| = E_p \pm \frac{\sqrt{3}a}{2} V_{pp\pi} |\Delta k|$ and eigenvectors $\begin{bmatrix} \pm e^{i\theta/2} \\ e^{-i\theta/2} \end{bmatrix}$

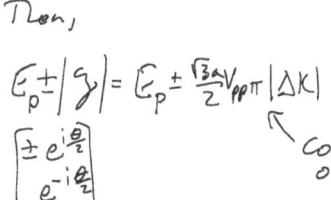

← Constant on circle

In the vicinity of the Dirac point, the spectrum looks like two cones:

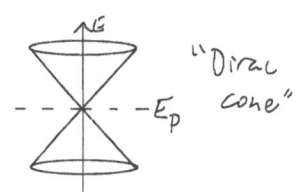

"Dirac cone"

Since $\mathcal{H} \propto \cos\theta \sigma_x + \sin\theta \sigma_y$, it is tempting to consider the orbital components from inequivalent lattice sites as 2-component spinor "pseudospin". However, it is completely unrelated to angular momentum and magnetic moment like true spin.

"Massless Dirac fermions"

The total energy from solution of the relativistic Dirac Eqn is $E = \pm\sqrt{(mc^2)^2 + (pc)^2}$. For small momentum p, this can be expanded as $\pm mc^2 \sqrt{1 + \left(\frac{pc}{mc^2}\right)^2} \sim \pm mc^2 \left(1 + \frac{p^2}{2m^2c^2} + \cdots\right) = \pm\left(mc^2 + \frac{p^2}{2m} + \cdots\right)$

This gives a dispersion

for $p = \hbar k$, $m \neq 0$ (like electrons):

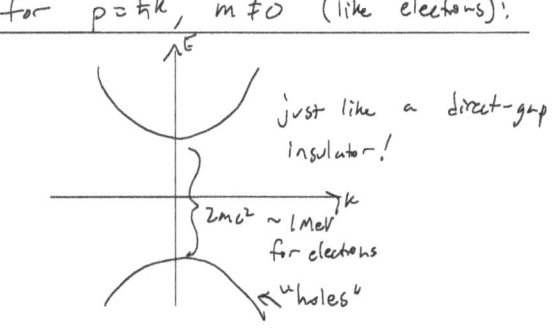

just like a direct-gap insulator!

$2mc^2 \sim 1 \text{MeV}$ for electrons

"holes"

for $m = 0$ (like photons) $E = \pm pc$:

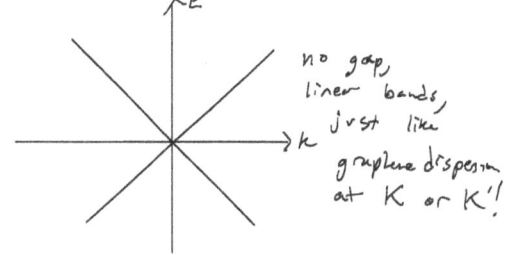

no gap, linear bands, just like graphene dispersion at K or K'!

For photons, $V_g = \frac{1}{\hbar} \nabla_k E = c$. For electrons in graphene, lattice const $a \sim 2.5 \text{Å}$ so

$V_g \sim \frac{1}{\hbar} \frac{10 \text{eV}}{\pi/a} \sim \frac{10 \text{eV} \cdot 2.5 \times 10^{-8} \text{cm}}{6.6 \times 10^{-16} \text{eV·s} \cdot \pi} \sim 10^8 \text{ cm/s}$ (about $c/100$)

3D LCAO: Cubic Lattice

"Recipe":
1. define geometry of lattice: determine nearest neighbor vectors
2. Identify all orbitals in primitive unit cell (basis set)
3. Determine matrix elements (with appropriate phases) for each orbital combination.

In 3D cubic lattice w/ s, p_x, p_y, p_z we have:

"diagonal" elements:

"off-diagonal" elements:

e.g. $\langle s|\mathcal{H}|p_y\rangle$
so $\langle s|\mathcal{H}|p_i\rangle = 0$ except for $\hat{x}\|\hat{i}$

e.g. $\langle p_y|\mathcal{H}|p_z\rangle$
so $\langle p_i|\mathcal{H}|p_j\rangle = 0$ for $i \neq j$

Hamiltonian matrix

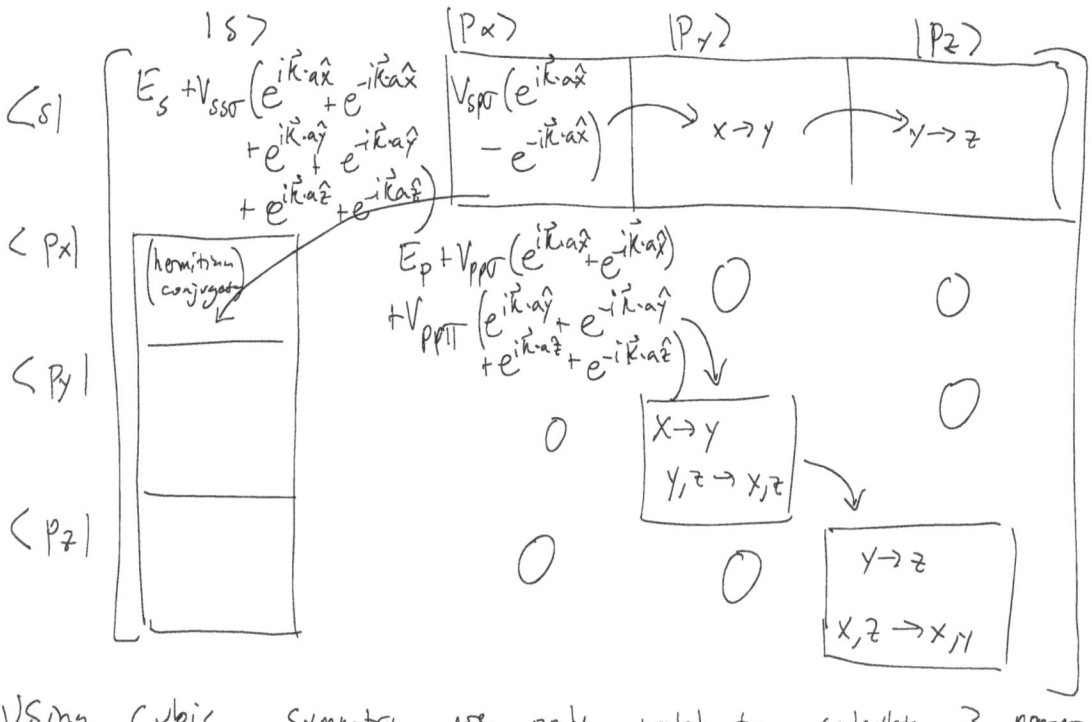

Using cubic symmetry, we only needed to calculate **3** nonzero elements!

Cubic Bandstructure in sp3 basis

The valence band is 3× degenerate @ Γ point, but these split into two bands for $k \neq 0$: light and heavy holes, resulting from the different kind of orbital overlap in different directions for nonisotropic p orbitals.

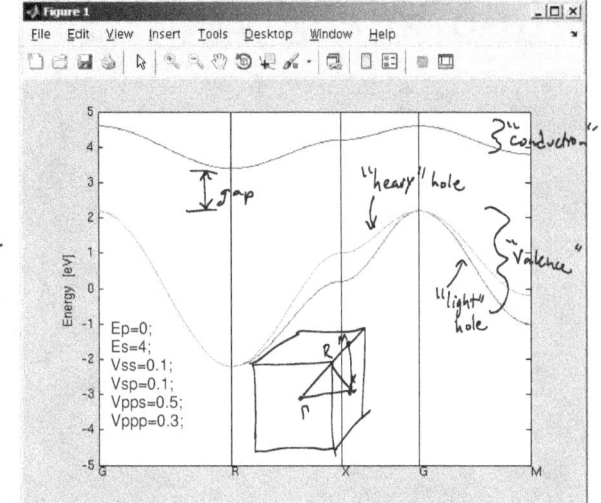

Note that the p-like valence band arises from atomic states w/ nonzero angular momentum $\ell = 1$. When spin-orbit interaction is included, the $j = \ell + \frac{1}{2} = \frac{3}{2}$ states ($m_J = \pm\frac{3}{2}, \pm\frac{1}{2}$) have different energy than the $j = \ell - \frac{1}{2} = \frac{1}{2}$ states ($m_J = \pm\frac{1}{2}$). This is the reason for the "split-off" valence band!

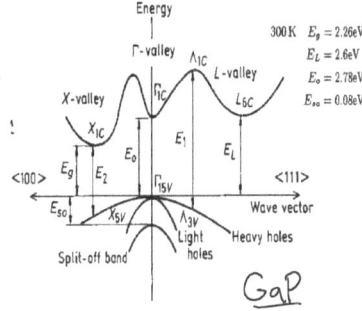

Spin-orbit interaction

Expand sp^3 orbital basis to $\{s\uparrow, s\downarrow, p_x\uparrow, p_x\downarrow, \text{etc}\}$

via $\mathcal{H} \otimes \mathbb{I}_2$ (an 8×8 matrix w/ 2-fold spin degeneracy)
   (4×4 orbital   (2×2 identity)
    hamiltonian)

"direct"/"kronecker" product

Now, include spin-orbit term $\propto \hat{L}\cdot\hat{S}$, nonzero only for the six $\ell=1$ p-states!

$\hat{L}\cdot\hat{S} = L_x \otimes S_x + L_y \otimes S_y + L_z \otimes S_z$ (a 6×6 matrix)

where $\hat{S} = \frac{\hbar}{2}[\sigma_x \hat{x} + \sigma_y \hat{y} + \sigma_z \hat{z}]$ (2×2 Pauli matrices)

and

$\hat{L}_z = \hbar \begin{bmatrix} 1 & 0 & 0 \\ 0 & 0 & 0 \\ 0 & 0 & -1 \end{bmatrix}$, $\hat{L}_y = \frac{\hbar}{\sqrt{2}}\begin{bmatrix} 0 & -i & 0 \\ i & 0 & -i \\ 0 & i & 0 \end{bmatrix}$, $\hat{L}_x = \frac{\hbar}{\sqrt{2}}\begin{bmatrix} 0 & 1 & 0 \\ 1 & 0 & 1 \\ 0 & 1 & 0 \end{bmatrix}$

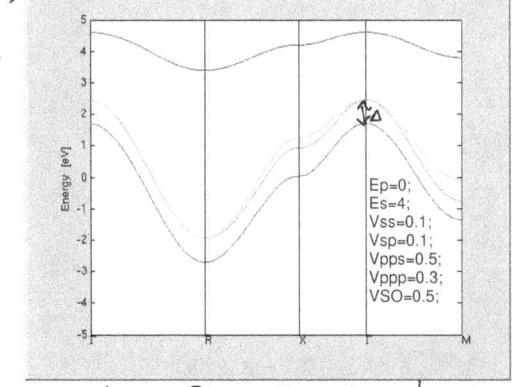

but this is in basis of definite $\langle L_z \rangle = \hbar m_\ell$, NOT in p_x, p_y, p_z basis!

So transform $\hat{L}\cdot\hat{S}$ to orbital basis

$p_x = \frac{1}{\sqrt{2}}(|m_\ell = -1\rangle - |m_\ell = +1\rangle)$ $p_y = \frac{i}{\sqrt{2}}(|m_\ell = -1\rangle + |m_\ell = +1\rangle)$

and add $\lambda \hat{L}\cdot\hat{S}$ to 6×6 p-state subspace of Hamiltonian, then diagonalize!

3D LCAO: non-cubic lattice (Slater + Koster, Phys Rev **94**, 1498 (1954))

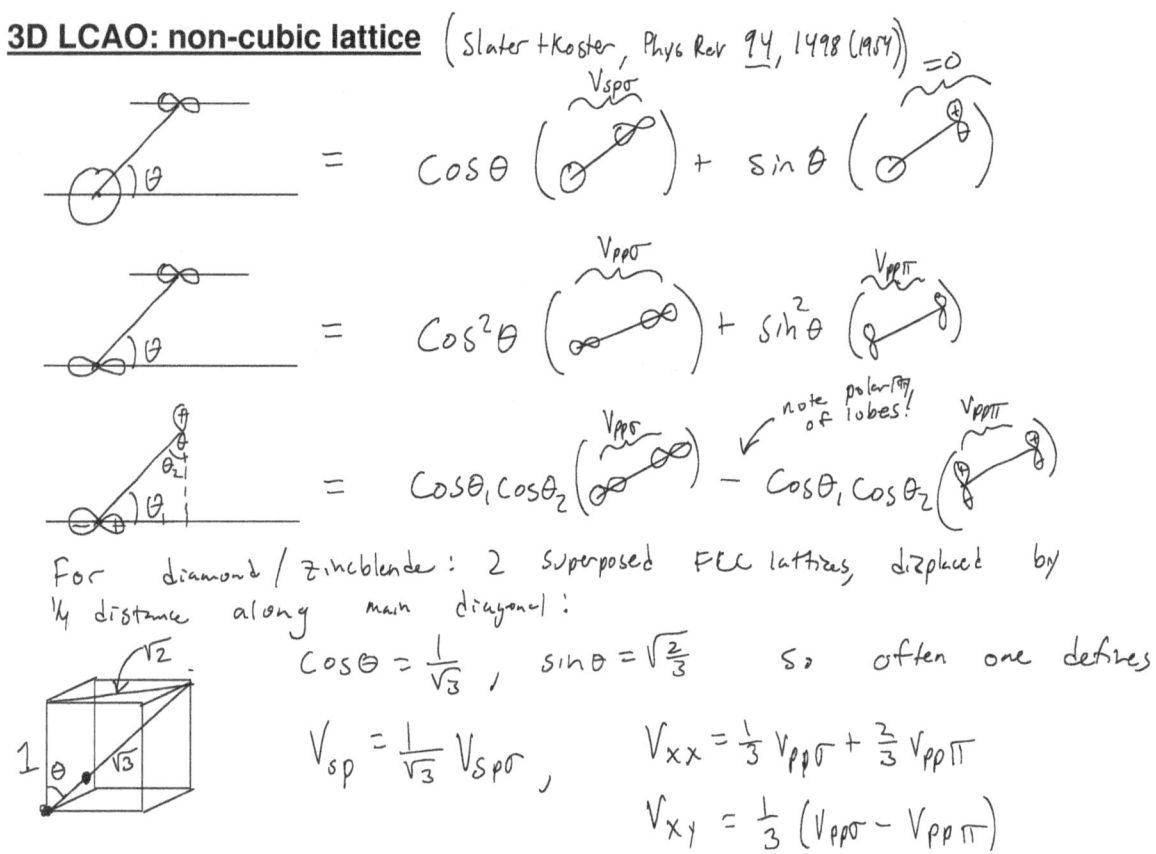

For diamond/zincblende: 2 superposed FCC lattices, displaced by 1/4 distance along main diagonal:

$\cos\theta = \frac{1}{\sqrt{3}}, \quad \sin\theta = \sqrt{\frac{2}{3}}$ so often one defines

$V_{sp} = \frac{1}{\sqrt{3}} V_{sp\sigma}, \quad V_{xx} = \frac{1}{3} V_{pp\sigma} + \frac{2}{3} V_{pp\pi}$

$V_{xy} = \frac{1}{3}(V_{pp\sigma} - V_{pp\pi})$

Carriers in Semiconductors

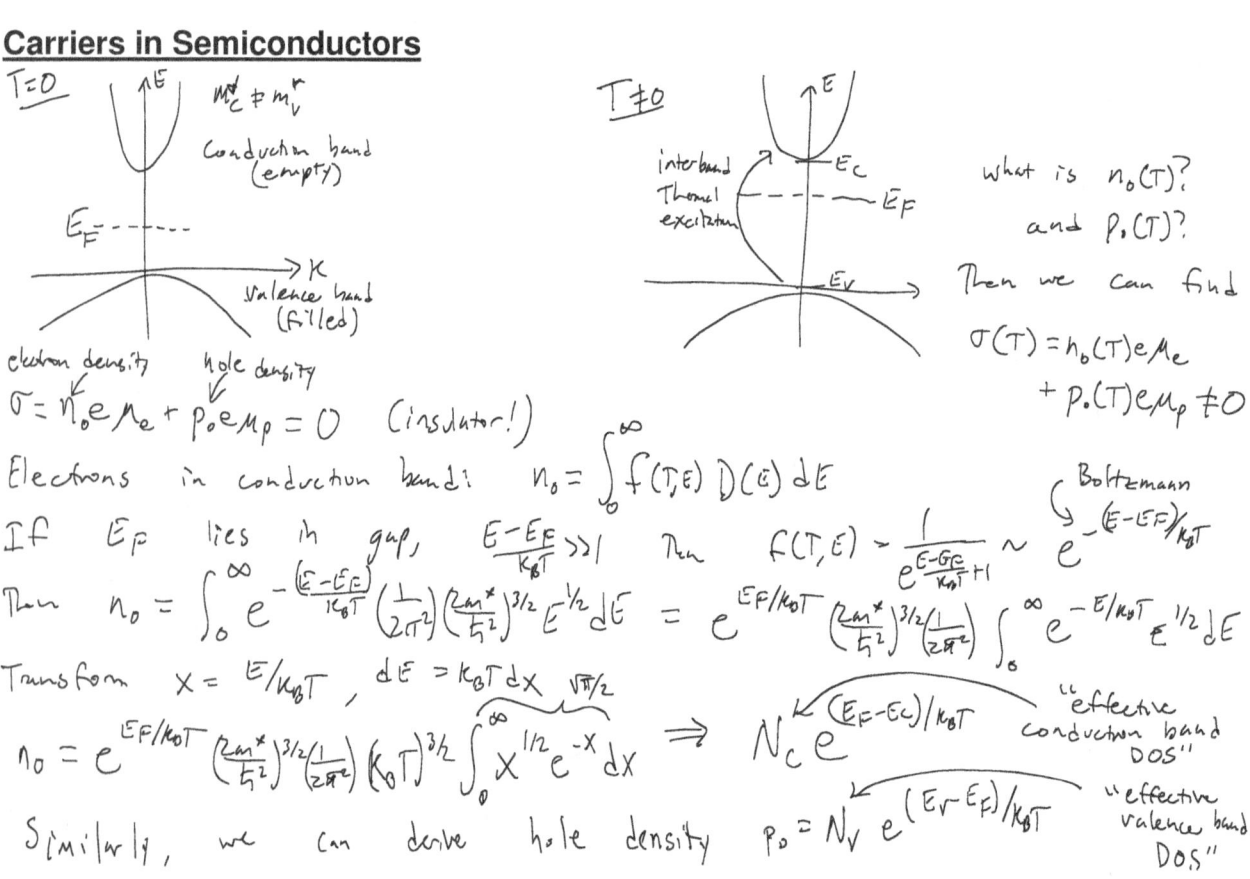

$\sigma = n_0 e \mu_e + p_0 e \mu_p = 0$ (insulator!) [electron density, hole density]

Electrons in conduction band: $n_0 = \int_0^\infty f(T,E) D(E) dE$

If E_F lies in gap, $\frac{E-E_F}{k_B T} \gg 1$ then $f(T,E) = \frac{1}{e^{(E-E_F)/k_B T}+1} \sim e^{-(E-E_F)/k_B T}$ (Boltzmann)

Then $n_0 = \int_0^\infty e^{-(E-E_F)/k_B T} \left(\frac{1}{2\pi^2}\right)\left(\frac{2m^*}{\hbar^2}\right)^{3/2} E^{1/2} dE = e^{E_F/k_B T} \left(\frac{2m^*}{\hbar^2}\right)^{3/2} \left(\frac{1}{2\pi^2}\right) \int_0^\infty e^{-E/k_B T} E^{1/2} dE$

Transform $x = E/k_B T$, $dE = k_B T dx$

$n_0 = e^{E_F/k_B T} \left(\frac{2m^*}{\hbar^2}\right)^{3/2} \left(\frac{1}{2\pi^2}\right) (k_B T)^{3/2} \underbrace{\int_0^\infty x^{1/2} e^{-x} dx}_{\sqrt{\pi}/2} \Rightarrow N_c e^{(E_F-E_c)/k_B T}$ "effective conduction band DOS"

Similarly, we can derive hole density $p_0 = N_v e^{(E_v-E_F)/k_B T}$ "effective valence band DOS"

Law of Mass Action

For "intrinsic" insulator/semiconductor $n_0 = p_0$

$$N_c e^{(E_F - E_C)/k_B T} = N_v e^{(E_V - E_F)/k_B T}$$

$$e^{(2E_F - (E_C + E_V))/k_B T} = \frac{N_v}{N_c} \Rightarrow e^{2(E_F - E_m)/k_B T} = \frac{N_v}{N_c}$$

$$E_F - E_m = \frac{k_B T}{2} \ln \frac{N_v}{N_c}$$

So as T rises, E_F moves toward band w/ lower m^*!

Note: we have reduced our system of a gapped parabolic dispersion to a simple (but highly degenerate) 2-level system!
$E_g \updownarrow$ ——— E_C
——— $E_m = \frac{E_C + E_V}{2}$
——— E_V

Note: $n_0 p_0$ is independent E_F

$$n_0 p_0 = N_c N_v e^{-(E_C - E_V)/k_B T} = N_c N_v e^{-E_g/k_B T} = n_i^2$$

$$n_0 p_0 = n_i^2 \quad \text{"Law of mass action"}$$
True even when $n_0 \neq p_0$! @RT

Example: In Si, $N_c = 3 \times 10^{19} \text{cm}^{-3}$ and $N_v = 2 \times 10^{19} \text{cm}^{-3}$, $E_g = 1.1 \text{eV}$, $k_B T = 1/40 \text{eV}$

$$n_i = \sqrt{N_c N_v} e^{-E_g/2k_B T} \sim 10^{19} \text{cm}^{-3} \cdot e^{-20} \sim 10^{10} \text{cm}^{-3}$$

intrinsic conductivity $\sigma_e = n_e e \mu_e \sim 10^{10} \text{cm}^{-3} \cdot 10^{-19} C \cdot 10^3 \frac{\text{cm}^2}{V \cdot s} = 10^{-6} \text{S cm}^{-1}$ @RT

Controlling E_F: from intrinsic to extrinsic

How to gain more of one carrier in one band at the expense of opposite carriers in other band?

<u>electrons</u>: "donor" impurity doping → "n-type"

Substitute atoms w/ more valence electrons: group V in place of group IV (Si, Ge, C)
(P, As, Sb)

for compound III-V, group IV in place of group III atom

<u>holes</u>: "acceptor" impurity doping → "p-type"

Substitute atoms w/ less valence electrons: group III in place of group IV (Si, Ge, C)
(B, Al, Ga)

for compound III-V, group II in place of group V atom (?)

Example: $n \approx N_D = N_c e^{(E_F - E_C)/k_B T}$ ← donor concentration
$\rightarrow E_F - E_C = k_B T \ln \frac{N_D}{N_c}$

For $N_D = 10^{15} \text{cm}^{-3}$ and $N_c \sim 10^{19} \text{cm}^{-3}$, $E_F - E_C \sim -0.26 \text{eV}$ @RT (closer to E_C than E_V if $E_g \sim 1 \text{eV}$)

This concentration gives $\rho = 1/\sigma = \frac{1}{n e \mu} \sim \frac{1}{10^{15} \text{cm}^{-3} \cdot 10^{-19} \cdot 10^3}^{(Si)} \sim 10 \, \Omega \text{cm}$

Impurity state

Donors leave behind positively charged immobile ion → hydrogenic spectrum

Bohr atom: $E_0 = -\frac{1}{2}\alpha^2 mc^2 = -\frac{1}{2}\left(\frac{e^2}{4\pi\epsilon_0 \hbar c}\right)^2 mc^2 = -13.6\,eV$

impurity atom in semiconductor:

$$E_0 = -\frac{1}{2}\left(\frac{e^2}{4\pi\epsilon_0 \hbar c}\right)^2 mc^2 \left(\frac{m^*}{m}\frac{1}{\epsilon_{rel}^2}\right) \xrightarrow{Si} -13.6\,eV\left(\frac{0.3}{12^2}\right) \sim -30\,meV$$

This is in the gap, close to E_c + easily ionized @ RT!

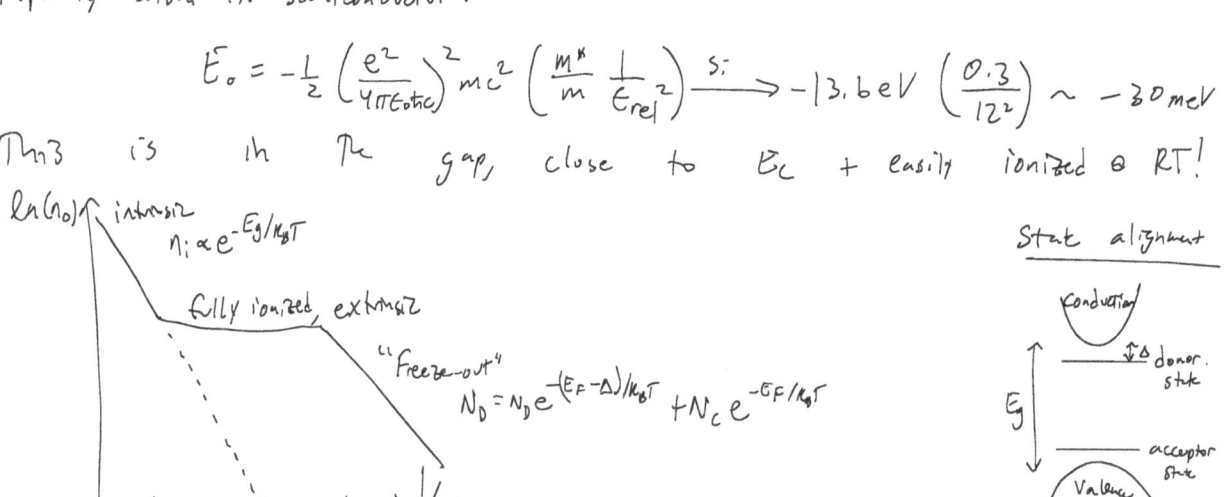

State alignment

Metal-Insulator transition

Can we dope to arbitrary concentration and still expect to see insulating behavior ("freezeout") at low enough temperatures? NO!

At high enough carrier concentration, screening will diminish the attractive impurity potential to the point that no bound state is supported!

A crude approximation of this density is provided by the 3D spherical finite well of depth V_0 and radius a_0, where $V_0 a_0^2 > \frac{\hbar^2 \pi^2}{8m}$ must be satisfied to support a bound state.

Thomas-Fermi screening length: $a_0^2 = \frac{\epsilon}{e^2 D(E_F)} = \frac{2\epsilon E_F}{3e^2 n} = \frac{2\epsilon \frac{\hbar^2}{2m}(3\pi^2 n)^{2/3}}{3e^2 n}$

So we must have

$$V_0 \frac{2\epsilon \frac{\hbar^2}{2m}(3\pi^2 n)^{2/3}}{3e^2 n} > \frac{\hbar^2 \pi^2}{8m} \rightarrow n^{1/3} < \frac{8V_0\epsilon}{e^2(3\pi^2)^{1/3}} = \frac{8}{(3\pi^2)^{1/3}}\frac{V_0\,\epsilon_{rel}}{\alpha \hbar c} \sim \frac{10^{-2}eV \cdot 10}{10^{-2}\,10^{-5}eV/cm}$$

$$= 10^6\,cm^{-1}$$

So for $N_D \gtrsim 10^{18}\,cm^{-3}$, we no longer expect insulating behavior → metal!

Band diagrams

Ability to dope semiconductor as a function of position allows for fabrication of useful devices! How to understand this spatial dependence + how it affects transport of charge?

Kroemer's "Lemma of Proven ignorance":

"If you can't draw energy-band diagram, you don't know what you're talking about" — H. Kroemer RMP 73, 783 (2001) (Nobel Prize, 2000)

<u>Band diagram</u>: band extrema [from $E(k)$] <u>and</u> potential energy [$(-e)\phi$] v.s. position

Energy = $E(k) + (-e)\phi$

homogeneous n-type, equilibrium

homogeneous p-type, equilibrium

<u>equilibrium</u> ≡ Fermi level is constant

What happens at an interface between semiconductors of different doping?

p-n junction in equilibrium

Charge transfer to equalize chemical potential results in dipole formation!

Poisson's eq.
$$-\nabla^2 \phi = \frac{\rho}{\epsilon}$$

⟹ constant space-charge density ⟹ quadratic $\phi(x)$

"built-in barrier" V_b metallurgical junction

"depletion regions" d_n, d_p

- total charge neutrality $N_D d_n = N_A d_p$ gives depletion region depths inversely prop. to doping density: eg. $d_n = \left(\frac{2\epsilon V_b}{e} \frac{N_A/N_D}{N_A + N_D} \right)^{1/2}$

- Internal electric field results from spatially distributed dipole.

- Equilibrium maintained by equalizing <u>drift</u> and <u>diffusion</u> currents

What happens out of equilibrium (voltage bias)?

Drift-Diffusion equation

Particle current $\quad J = \overbrace{nv}^{\text{drift}} - \overbrace{D\nabla n}^{\text{diffusion}}$

Continuity equation $\quad \frac{dn}{dt} = -\nabla\cdot J - \frac{n}{\tau}\leftarrow$ "minority carrier lifetime"

The $-\frac{n}{\tau}$ term is due to "recombination": annihilation of electron by hole via transition across gap. In direct bandgap, energy conservation can result in emission of photon w/ $\hbar\omega \sim E_g$.

Combining two above gives:

$\frac{dn}{dt} = D\nabla^2 n - v\cdot\nabla n - \frac{n}{\tau} \quad$ "drift-diffusion equation" $\rightarrow$ governs spatio-temporal evolution of excess carriers!

In steady-state, $\frac{dn}{dt} = 0$. For $v=0$, we have (in 1-D)

$\frac{d^2 n}{dx^2} = \frac{n}{D\tau} \implies n(x) = C_+ e^{+x/L} + C_- e^{-x/L}, \quad L = \sqrt{D\tau} \quad$ "diffusion length"

The coefs $C_{+/-}$ determined by B.C.'s. What are the B.C.'s for p-n under bias?

p-n junction in forward bias

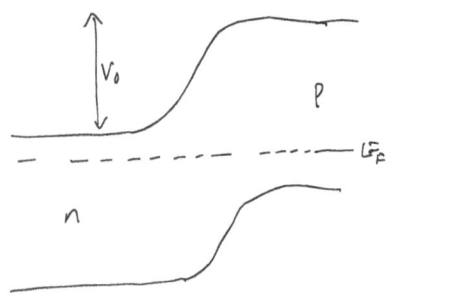

Lowering barrier by amount eU allows high concentration of carriers closer to E_F to be capable of diffusion across depletion region. Determines B.C. since excess electrons $\Delta n = n_p e^{eU/kT} - n_p$ @ start of p depletion.

Our solution is then

$n(x) = n_p(e^{eU/kT} - 1) e^{-x/L} \quad$ "minority carrier injection"

(Similarly for holes moving in opposite direction)

"quasi Fermi levels"

linear since $F_n - E_F \propto \ln\frac{1}{n_0}$

Ideal diode equation

Diffusion current from $n \to p$ is $J_D = -D\frac{dn}{dx} = \frac{D}{L}n_p(e^{eU/kT}-1)$

Same analysis for holes gives total charge current

$$J_c = e\left(\frac{D_n}{L_n}n_p + \frac{D_p}{L_p}p_n\right)(e^{eU/kT}-1) \quad \text{"ideal diode equation"}$$

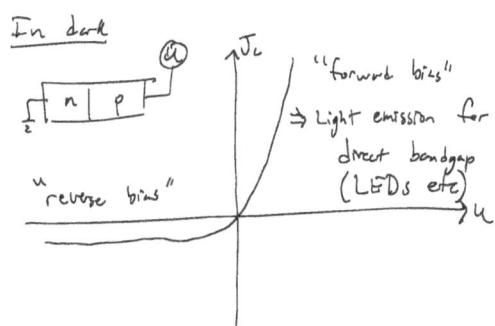

In dark

"forward bias" ⇒ Light emission for direct bandgap (LEDs etc)

"reverse bias"

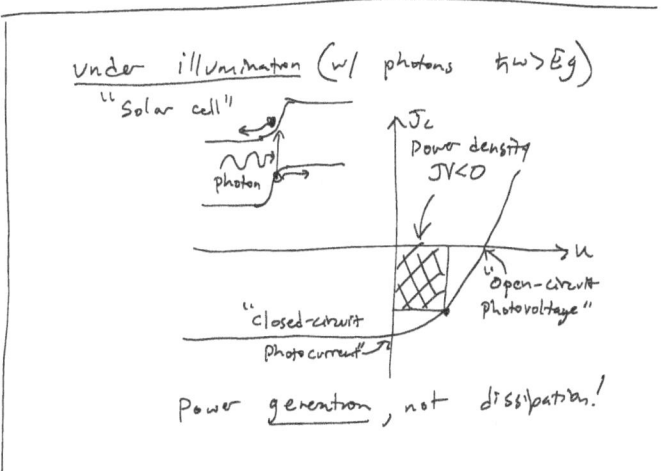

Under illumination (w/ photons $\hbar\omega > E_g$)
"Solar cell"

Power density $JV < 0$

"closed-circuit photocurrent"   "Open-circuit photovoltage"

Power generation, not dissipation!

lower barrier: "forward bias": exponential, dominated by diffusion current

higher barrier: "reverse bias" saturated, dominated by drift current and limited by thermal generation

A quick tour of semiconductor devices based on the p-n junction

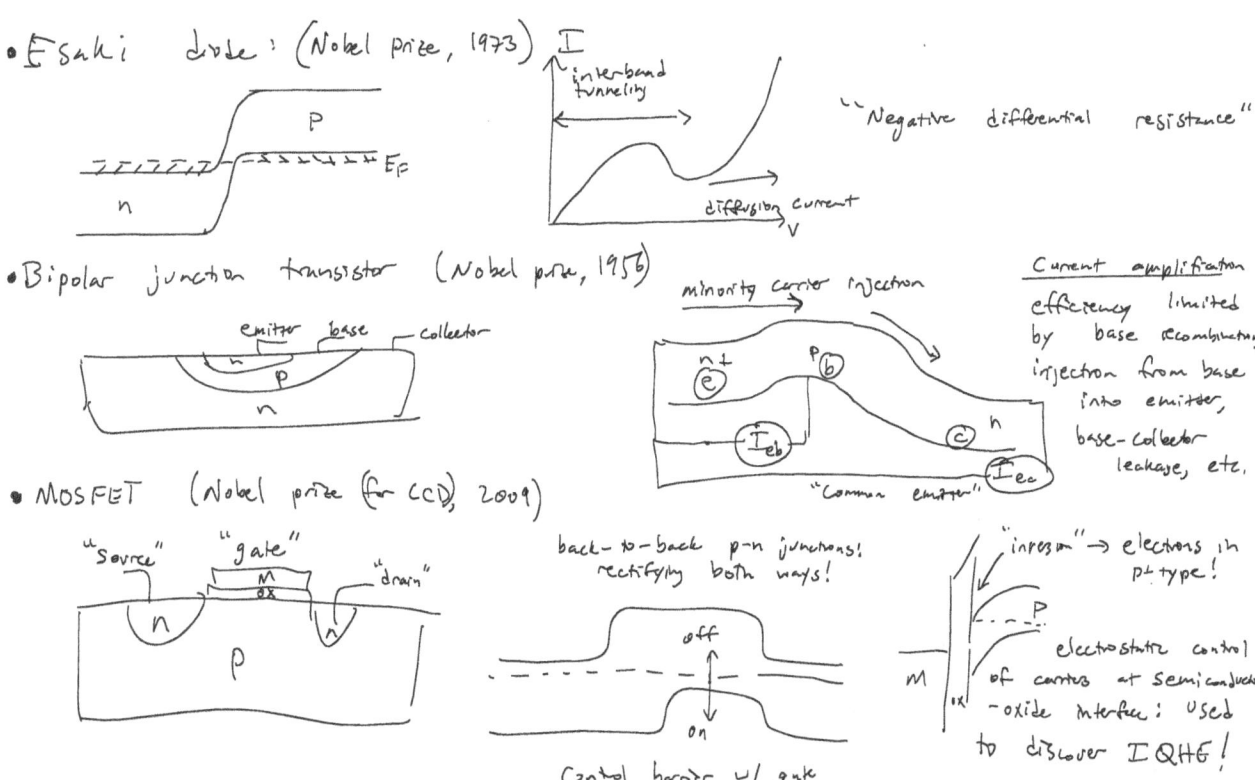

- Esaki diode: (Nobel prize, 1973)
  interband tunneling
  "Negative differential resistance"
  diffusion current

- Bipolar junction transistor (Nobel prize, 1956)
  emitter, base, collector
  minority carrier injection
  "Common emitter"

  Current amplification efficiency limited by base recombination, injection from base into emitter, base-collector leakage, etc.

- MOSFET (Nobel prize (for CCD) 2009)
  "Source" "gate" "drain"
  back-to-back p-n junctions! rectifying both ways!
  off / on
  Control barrier w/ gate

  "inversion" → electrons in p-type!
  electrostatic control of carriers at semiconductor-oxide interface: used to discover IQHE!

Semiconductor

2DEG band diagram:

Confinement energy can be substantially larger than $k_B T$, allowing population solely of ground-state "sub-band".

By providing carriers to the well without putting ionized impurities there, one can obtain 2DEG mobilities $\mu > 50 \times 10^6 \frac{cm^2}{Vs}$!

$\Rightarrow$ used to discover FQHE, and to observe QPC, etc via electrostatic top-gate depletion

Homework 1:

1. Justify the definition of "cyclotron" frequency $\omega_p = eB/m$ by showing that the solutions of the coupled x- and y- equations of motion in the absence of an E-field and without scattering (only Lorentz force) describe circular trajectories.

2. In the Drude model, scattering is inelastic, i.e. energy is lost. First, derive the energy lost per collision. Then, by averaging over the probability distribution of scattering times, $\frac{e^{-t/\tau}}{\tau}$, derive the power density lost to the ions by collisions. Show that this "Joule heating" is consistent with the geometry-dependent expression $P = I^2 R$.

3. Propagation of EM waves in metals:
   a. We used a high-frequency approximation to the complex AC conductivity in the derivation of the optical surface reflection coefficient R. Using a numerical program such as MATLAB, calculate and plot R through the plasmon frequency range without making this approximation. Use several values of τ in the range $1 fs < \tau < 100 fs$, and investigate the role of static permittivity ε and electron density n.
   b. The transmitted wave will be attenuated. Find the penetration or "skin" depth and plot it as a function of ω along with R above.

4. Thermoelectric efficiency is often determined by the dimensionless figure of merit ZT, where $Z = \frac{\sigma K^2}{\kappa}$. Explicitly verify its dimensionality and justify the use of this expression to maximize thermoelectric effects.

5. Helicon Waves
   Assume a magnetic field **H** along the z-axis, and a circularly polarized electric-field plane wave ($E_y = \pm i E_x$) propagating along the same direction in a Drude metal.
   a. Generalize the AC Drude conductivity to include the effect of the magnetic field.
   b. Determine the corresponding frequency-dependent effective permittivity.
   c. Assuming $\omega_c \tau \gg 1$, sketch ε for $\omega > 0$ and demonstrate that solutions to the dispersion relation $k^2 c^2 = \varepsilon \omega^2$ exist for all k.
   d. Show that when $\omega \ll \omega_c$, the dispersion relation is $\omega = \omega_c (k^2 c^2 / \omega_p^2)$. This is the *helicon wave*.
   e. Estimate the 1-cm wavelength helicon frequency in a typical metal at $|H| = 1T$.

6. Surface Plasmons
   An electromagnetic wave that can propagate along the surface of a metal complicates the observation of ordinary (bulk) plasmons. Let the metal be contained in the half space $z > 0$, with the region $z < 0$ being vacuum. Assume that the bulk electric charge

density r appearing in Maxwell's equations vanishes both inside and outside the metal (though a surface charge density can appear in the plane z = 0). The surface plasmon is a solution to Maxwell's equations where the time dependence is of the form e-iωt and the space dependence is of the form:

E_x = A exp(iqx)exp(-Kz); E_y = 0; E_z = B exp(iqx)exp(-Kz), (z > 0);
E_x = C exp(iqx)exp(K'z); E_y = 0; E_z = D exp(iqx)exp(K'z) (z < 0);

where q is real and K and K' both real and positive. (Note that the amplitude decays exponentially away from the surface both for z positive and negative.)

Maxwell's equations imply that

$$\vec{\nabla} \cdot (\epsilon \vec{E}) = 0,$$

$$-\nabla^2 \mathbf{E} = \left(\frac{\omega}{c}\right)^2 \epsilon(\omega) \mathbf{E},$$

where ε(ω) is the frequency dependent dielectric constant, and are to be solved with the usual boundary conditions, ($E_\parallel$ continuous, (εE)⊥ continuous). (At the low-q region of interest here, the longitudinal and transverse dielectric constants are equal.) Assume the Drude result

$$\sigma(\omega) = \frac{\sigma_0}{1 - i\omega\tau}, \qquad \sigma_0 = \frac{ne^2\tau}{m},$$

and note the general relation between conductivity and dielectric constant.

a. Find three equations relating q, K, and K' as functions of ω. Note: The coefficients A, B, C and D should not appear in these equations.
b. Hence, eliminating K and K', show that $\omega^2 = c^2 q^2 [1 + 1/\epsilon(\omega)]$
c. For the high-frequency asymptotic form of ε(ω), sketch $c^2 q^2$ as a function of ω
d. Hence sketch ω as a function of q.
e. In the limit as cq >> ω, show that there is a solution at frequency $\omega_p/\sqrt{2}$, where ω_p is the bulk plasmon frequency. What is ε(ω) at this frequency? Show from an examination of K and K' that the wave is confined to the surface. Describe its polarization. This wave is known as a *surface plasmon*.
f. Show that there are also solutions with ω > ω_p. However, these are unphysical. Why?

1) $\frac{d\vec{p}}{dt} = -e\left(\frac{\vec{p}}{m} \times \vec{B}\right)$

for $\vec{B} = B_z \hat{z}$, we have only 2 nonzero components:

$\dot{p}_x = -\frac{eB_z}{m} p_y$ \quad (1)

$\dot{p}_y = \frac{eB_z}{m} p_x$ \quad (2)

From (2), we have $\ddot{p}_y = \frac{eB_z}{m} \dot{p}_x = -\left(\frac{eB_z}{m}\right)^2 p_y = -\omega_c^2 p_y$

Solution $p_y(t) = A\cos\omega_c t$ \quad so \quad $y(t) = \frac{A}{m\omega_c}\sin\omega_c t$

Then (1) gives $p_x = -A\sin\omega_c t$ \quad so \quad $x(t) = \frac{A}{m\omega_c}\cos\omega_c t$

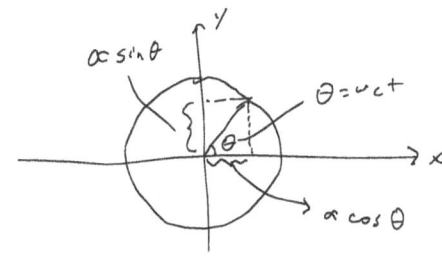

2) Energy lost per collision after time t:

$\Delta E = \frac{1}{2}m\left(\frac{e\mathcal{E}t}{m}\right)^2 = \frac{e^2\mathcal{E}^2}{2m}t^2$

$\langle \Delta E \rangle = \int_0^\infty \Delta E \frac{e^{-t/\tau}}{\tau} dt = \frac{e^2\mathcal{E}^2}{2m\tau}\int_0^\infty t^2 e^{-t/\tau} dt$

By e.g. differentiation under the integral,

$\int_0^\infty t^2 e^{-t/\tau} dt = \frac{d^2}{d(\frac{1}{\tau})^2}\int_0^\infty e^{-t/\tau} dt = \frac{d^2}{d(\frac{1}{\tau})^2}\left[-\tau e^{-t/\tau}\Big|_0^\infty\right] = \frac{d^2}{d(\frac{1}{\tau})^2}\tau = 2\tau^3$

So $\langle \Delta E \rangle = \frac{e^2\mathcal{E}^2}{2m\tau} \cdot 2\tau^3 = \frac{e^2\mathcal{E}^2\tau^2}{m}$ \quad and power dissipated per volume

$P = \frac{\langle \Delta E \rangle}{\tau} = \left(\frac{e^2\tau}{m}\mathcal{E}\right)\mathcal{E} = \vec{j}\cdot\vec{\mathcal{E}}$

This is the same as $P = IV = j \cdot A \mathcal{E} L = \vec{j}\cdot\vec{\mathcal{E}} \cdot \text{Volume}$

3) Transmitted plane waves $e^{i(kx-\omega t)}$ have attenuation length

$$\left(\text{Im}\{k\}\right)^{-1} = \left(\text{Im}\left\{\frac{\omega\sqrt{\epsilon_{rel}}}{c}\right\}\right)^{-1}$$

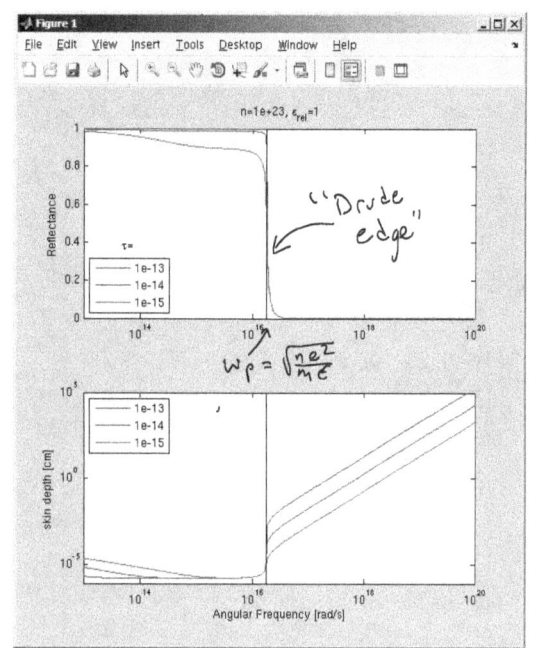
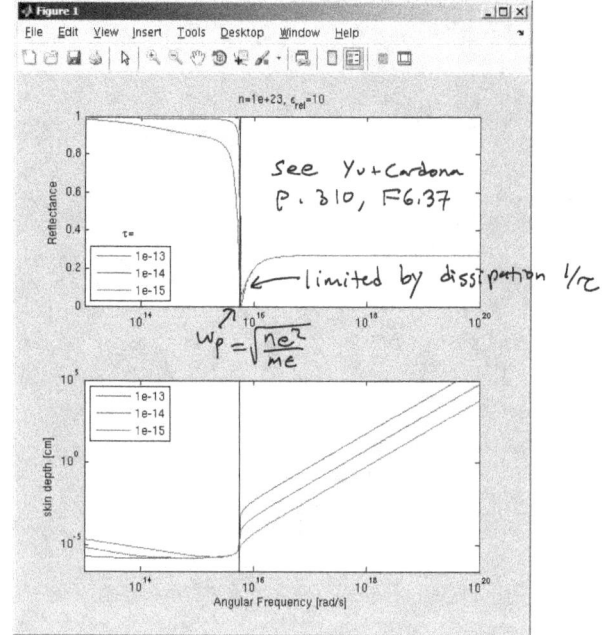

4) $Z = \frac{\sigma}{\kappa} Q^2$

we have seen that $\frac{\kappa}{\sigma} = LT$, where $[L] = \frac{V^2}{K^2}$, so $\left[\frac{\sigma}{\kappa}\right] = \frac{K}{V^2}$

Since $Q = \frac{\mathcal{E}}{\nabla T}$, $[Q] = \frac{V}{K}$ and $[Z] = \left[\frac{\sigma}{\kappa} Q^2\right] = \frac{1}{K}$.

Therefore, ZT is dimensionless.

- It is obvious that we want a large thermopower to convert ∇T into usable electrical power.

- Smaller thermal conductivity $\kappa = \frac{j_Q}{\nabla T}$ means less power lost to increasing "cold" sink entropy at fixed temperature gradient

5) Helicon waves

   (a) Eq^n of motion: $\dot{\vec{p}} = -e\left(\vec{\mathcal{E}} + \frac{\vec{p}}{m} \times \vec{B}\right) - \frac{\vec{p}}{\tau}$

   In freq domain, $-i\omega \vec{p}(\omega) = -e\vec{\mathcal{E}}(\omega) - \frac{e\vec{p}(\omega) \times \vec{B}}{m} - \frac{\vec{p}(\omega)}{\tau}$

   Vector components:

   $x:\quad -i\omega p_x = -e\mathcal{E}_x - \frac{e}{m}(p_y B_z - p_z \cancel{B_y}^{\;0}) - \frac{p_x}{\tau}$

   $y:\quad -i\omega p_y = -e\mathcal{E}_y - \frac{e}{m}(p_z \cancel{B_x}^{\;0} - p_x B_z) - \frac{p_y}{\tau}$

   Defining $\omega_c \equiv \frac{eB_z}{m}$, we have two coupled eqns

   $$-i\omega p_x = -e\mathcal{E}_x - \omega_c p_y - \frac{p_x}{\tau}$$
   $$-i\omega p_y = -e\mathcal{E}_y + \omega_c p_x - \frac{p_y}{\tau}$$

   Solving for p_y and substituting:

   $$-i\omega p_x = -e\mathcal{E}_x - \omega_c \left(\frac{-e\mathcal{E}_y + \omega_c p_x}{-i\omega + \frac{1}{\tau}}\right) - \frac{p_x}{\tau}$$

   If $\mathcal{E}_y = \pm i\mathcal{E}_x$

   $$p_x = \frac{-e\mathcal{E}_x \pm \frac{ei\omega_c \mathcal{E}_x}{-i\omega + 1/\tau}}{-i\omega + \frac{\omega_c^2}{-i\omega + 1/\tau} + \frac{1}{\tau}}$$

   multiply top and bottom by $-i\omega + \frac{1}{\tau}$ and factor

   $$p_x = \frac{-e\left((-i\omega + 1/\tau) \mp i\omega_c\right)}{\left((-i\omega + 1/\tau) + i\omega_c\right)\left((-i\omega + 1/\tau) - i\omega_c\right)}$$

   $$= \frac{-e\tau \mathcal{E}_x}{1 - i(\omega \mp \omega_c)\tau}$$

   Then, $\sigma = ne\mu = ne\left(\frac{p}{m\mathcal{E}}\right) = \frac{\left(\frac{ne^2\tau}{m}\right)}{1 - i(\omega \mp \omega_c)\tau} = \frac{\sigma_0}{1 - i(\omega \mp \omega_c)\tau}$

ⓑ Wave eqn becomes
$$-\nabla^2 \mathcal{E} = -\frac{\partial}{\partial t}\left(\mu \frac{\sigma_0 \mathcal{E}}{1-i(\omega\mp\omega_c)\tau} + \mu\epsilon\frac{\partial \mathcal{E}}{\partial t}\right)$$

For plane-wave solns $\mathcal{E} \propto e^{i(kz-\omega t)}$,

$$k^2 \mathcal{E} = \frac{i\mu\sigma_0 \omega \mathcal{E}}{1-i(\omega\mp\omega_c)\tau} + \frac{\omega^2}{c^2}\mathcal{E}$$

$$k^2 c^2 = \omega^2 + \frac{i\mu\sigma_0 \omega c^2}{1-i(\omega\mp\omega_c)\tau} = \omega^2 - \frac{\frac{\sigma_0}{\epsilon\tau}\omega}{1/\tau + (\omega\mp\omega_c)}$$

$$= \omega^2\left(1 - \frac{\omega_p^2}{\omega(\omega\mp\omega_c+i/\tau)}\right) = \epsilon\omega^2$$

Therefore $\epsilon(\omega) = 1 - \dfrac{\omega_p^2}{\omega(\omega\mp\omega_c+i/\tau)}$

ⓒ $\epsilon(\omega) \sim 1 - \dfrac{\omega_p^2}{\omega(\omega\mp\omega_c)}$

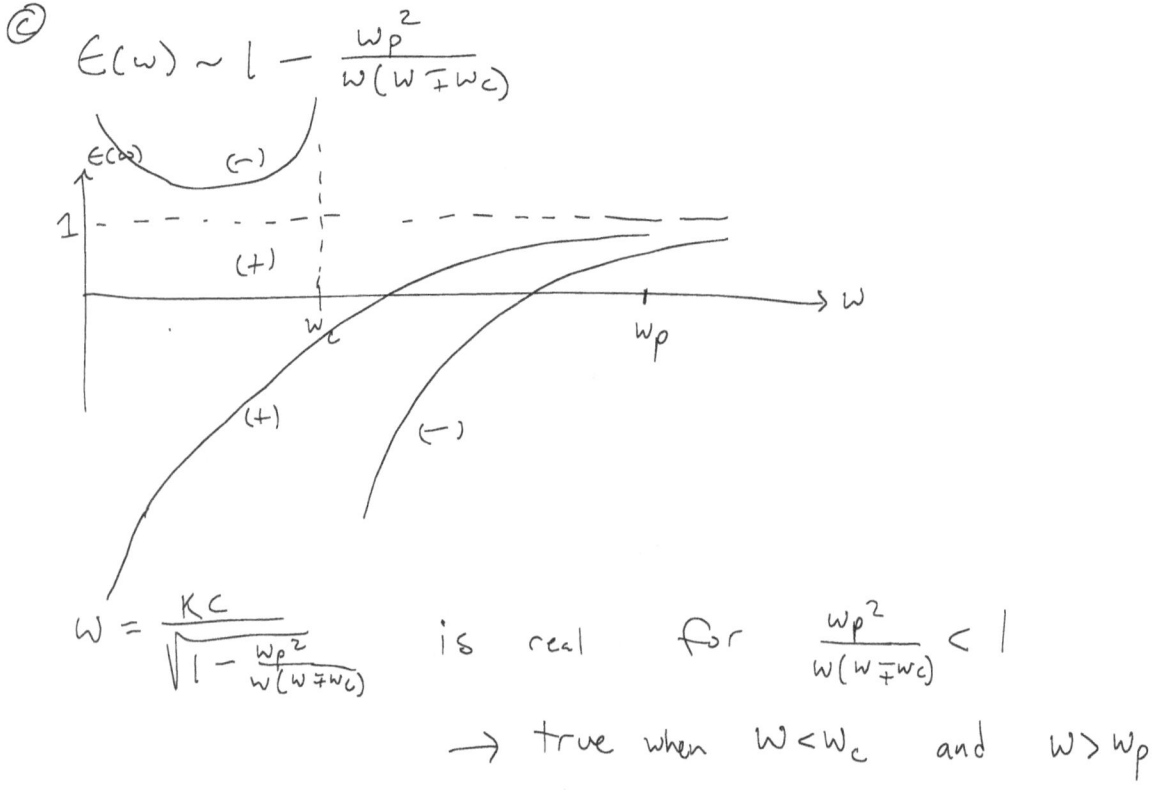

$\omega = \dfrac{kc}{\sqrt{1 - \frac{\omega_p^2}{\omega(\omega\mp\omega_c)}}}$ is real for $\dfrac{\omega_p^2}{\omega(\omega\mp\omega_c)} < 1$

→ true when $\omega < \omega_c$ and $\omega > \omega_p$

d)

$$\omega^2 = \frac{k^2c^2}{1 - \frac{\omega_p^2}{\omega(\omega \mp \omega_c)}} \quad \Rightarrow \quad \omega^2 - \frac{\omega^2 \omega_p^2}{\omega(\omega \mp \omega_c)} = k^2 c^2$$

For $\omega \ll \omega_c$, we can disregard terms $\propto \omega^2$

$$\frac{\omega \cdot \omega_p^2}{\omega_c} = k^2 c^2 \quad \longrightarrow \quad \omega = \omega_c \frac{k^2 c^2}{\omega_p^2}$$

e)

$$\omega_c = \frac{eB}{m} = \frac{2}{\hbar} \frac{e\hbar}{2m} B = 2 \frac{\mu_B B}{\hbar} \sim 2 \frac{5.8 \times 10^{-5} \, eV/T \cdot 1T}{6.6 \times 10^{-16} \, eV \cdot s} \sim 2 \times 10^{11} \, s^{-1}$$

$$\omega_p = \sqrt{\frac{ne^2}{m\epsilon}} \sim 10^{16} \, s^{-1}$$

So

$$\omega(1T) = 2 \times 10^{11} \, s^{-1} \left(\frac{\left(\frac{2\pi}{\lambda}\right)^2 c^2}{(10^{16} s^{-1})^2} \right) \approx 2 \times 10^{11} \frac{(2\pi)^2 \cdot 9 \times 10^{20} s^{-2}}{10^{32} s^{-2}}$$

$$\sim 60 \, Hz$$

6) Surface Plasmon

metal $\quad \rho = 0 \quad \epsilon(\omega) \quad\quad \mathcal{E}_x = A e^{iqx} e^{-kz}, \; \mathcal{E}_z = B e^{iqx} e^{-kz}$

vacuum $\quad \rho = 0 \quad \epsilon_0 \quad\quad \mathcal{E}_x = C e^{iqx} e^{k'z}, \; \mathcal{E}_z = D e^{iqx} e^{k'z}$

$$\vec{\nabla} \cdot (\epsilon \vec{\mathcal{E}}) = 0, \quad -\nabla^2 \vec{\mathcal{E}} = \left(\frac{\omega}{c}\right)^2 \epsilon \vec{\mathcal{E}}, \quad \sigma = \frac{\sigma_0}{1 - i\omega\tau} \quad \sigma_0 = \frac{ne^2\tau}{m}$$

$$\epsilon(\omega) = \epsilon \left(1 + \frac{i\sigma}{\epsilon\omega}\right)$$

Boundary conditions:

$\mathcal{E}_\parallel, \; \epsilon \mathcal{E}_\perp$ continuous

$A e^{iqx} = C e^{iqx} \quad\quad\quad B e^{iqx} \epsilon = D e^{iqx}$

$A = C \quad \text{①} \quad\quad\quad B = D/\epsilon \quad \text{②}$

Gauss' law $(\vec{\nabla}\cdot\vec{\mathcal{E}}=0)$

in metal ($z>0$) $\rightarrow$ $iq\mathcal{E}_x - K\mathcal{E}_z = 0$ $\rightarrow$ $iqA = KB$ ③

in air ($z<0$) $\rightarrow$ $iq\mathcal{E}_x + K'\mathcal{E}_z = 0$ $\rightarrow$ $iqC = -K'D$ ④

Wave eqn (Faraday + Ampere's)

in metal ($z>0$) $\rightarrow$ $(-q^2 + K^2)\mathcal{E} = -\frac{\omega^2}{c^2}\epsilon\mathcal{E}$ ⑤

in air ($z<0$) $\rightarrow$ $(-q^2 + K'^2)\mathcal{E} = -\frac{\omega^2}{c^2}\mathcal{E}$ ⑥

---

①,②,③,④: $\quad iqC = KB = -K'D = \frac{KD}{\epsilon}$

$\rightarrow K' = -\frac{K}{\epsilon}$

⑤,⑥: $\quad K^2 - K'^2 = \frac{\omega^2}{c^2}(1-\epsilon)$

$K^2 - \frac{K^2}{\epsilon^2} = \frac{\omega^2}{c^2}(1-\epsilon)$

$K^2 = \frac{\omega^2}{c^2}\frac{(1-\epsilon)}{1-\frac{1}{\epsilon^2}} = \frac{\omega^2}{c^2}\frac{\epsilon^2(1-\epsilon)}{\epsilon^2-1} = -\frac{\omega^2}{c^2}\frac{\epsilon^2}{1+\epsilon}$

$K'^2 = K^2 - \frac{\omega^2}{c^2}(1-\epsilon) = -\frac{\omega^2}{c^2}\left(\frac{\epsilon^2}{1+\epsilon} + 1 - \epsilon\right)$

$= -\frac{\omega^2}{c^2}\left(\frac{\epsilon^2 + 1 - \epsilon^2}{1+\epsilon}\right)$

If $\epsilon < -1$, $K = -\frac{\omega\epsilon}{c}\sqrt{\frac{-1}{1+\epsilon}}$, $K' = \frac{\omega}{c}\sqrt{\frac{-1}{1+\epsilon}}$

ⓑ: $(-q^2 + k'^2)\varepsilon = -\frac{\omega^2}{c^2}\varepsilon$ gives

$$q^2 = k'^2 + \frac{\omega^2}{c^2} = -\frac{\omega^2}{c^2}\left(\frac{1}{1+\epsilon} - 1\right)$$

$$= \frac{\omega^2}{c^2}\frac{\epsilon}{1+\epsilon}$$

So $\omega^2 = q^2 c^2 \left(1 + \frac{1}{\epsilon}\right)$

ⓒ $\omega\tau \gg 1$: $\sigma \to \frac{i\sigma_0}{\omega\tau}$, $\epsilon \to 1 - \frac{\omega_p^2}{\omega^2}$

$$\omega = qc\left(1 + \frac{1}{1 - \frac{\omega_p^2}{\omega^2}}\right)^{1/2}$$

[graph: qc vs ω, with asymptote at $\frac{\omega_p}{\sqrt{2}}$ and ω_p marked]

ⓓ

[graph: ω vs qc, with "continuum", "light like" $\omega = qc$, ω_p, $\frac{\omega_p}{\sqrt{2}}$ marked]

$$1 + \frac{1}{1 - \frac{\omega_p^2}{\omega^2}} = 0$$

$$\frac{\omega_p^2}{\omega^2} = 2$$

$$\omega = \frac{\omega_p}{\sqrt{2}}$$

ⓔ $\epsilon = 1 - \frac{\omega_p^2}{\omega^2} \sim -1$

$$k = -\frac{\omega\epsilon}{c}\sqrt{\frac{-1}{1+\epsilon}}$$

$$k' = \frac{\omega}{c}\sqrt{\frac{-1}{1+\epsilon}}$$

This is point at which k, k' become imaginary. Wave is bound for $\epsilon < -1$.

(e) slightly below $\omega = \frac{\omega_p}{\sqrt{2}}$, $\omega^2 = \frac{\omega_p^2}{2} - \delta$.

Then $\epsilon \sim 1 - \frac{\omega_p^2}{\frac{\omega_p^2}{2} - \delta} = 1 - \frac{2}{1 - \frac{2\delta}{\omega_p^2}} \sim 1 - 2\left(1 + \frac{2\delta}{\omega_p^2}\right)$

$$= -1 - \frac{4\delta}{\omega_p^2}$$

$\omega^2 = q^2 c^2 \left(1 + \frac{1}{\epsilon}\right) \Rightarrow q^2 c^2 = \frac{\omega^2}{1 + \frac{1}{\epsilon}} = \frac{\frac{\omega_p^2}{2} - \delta}{1 + \frac{1}{-1 - \frac{4\delta}{\omega_p^2}}}$

$$\simeq \frac{\frac{\omega_p^2}{2} - \delta}{1 - \left(1 - \frac{4\delta}{\omega_p^2}\right)} = \frac{\frac{\omega_p^4}{2} - \omega_p^2 \delta}{4\delta}$$

$$q^2 c^2 \sim \frac{\omega_p^4}{8\delta}$$

$iqA = KB$ ③

$iqC = -K'D$ ④

$K = -\frac{\omega \epsilon}{c}\sqrt{\frac{-1}{1+\epsilon}}$, $K' = \frac{\omega}{c}\sqrt{\frac{-1}{1+\epsilon}}$, $\omega^2 = q^2 c^2 \left(1 + \frac{1}{\epsilon}\right)$

$q = \frac{\omega}{c}\sqrt{\frac{\epsilon}{1+\epsilon}} = -\frac{K}{\sqrt{\epsilon}}$

$\frac{A}{B} = \frac{-iK}{q} = -i\sqrt{\epsilon}$ (elliptical)

$\frac{C}{D} = i\frac{K'}{q} = i$ (circular)

Homework 2:

1. Thermionic emission: use the following saturation current data to extract the cathode work function (in units of eV):

Temp [K]	current [A]
1200	3.53e-09
1250	1.54e-08
1300	7.17e-08
1350	2.82e-07
1400	1.33e-06
1450	3.48e-06
1500	1.15e-05

   What is the uncertainty in your calculation?

2. Fuchs-Sondheimer model: We used the Boltzmann Transport equation to derive the bulk conductivity of the free electron gas. A *thin film* is confined by its thickness t in the z-direction and so the occupancy distribution correction $g(k)$ is also z-dependent.
   a. By evaluating the *total* derivative dg/dt in the absence of translational symmetry in the z-dir, determine the correction to the relaxation term and write down the new Boltzmann equation for electric field in the x-dir.
   b. Solve this differential equation and write down two solutions: $g^+(k)$ for positive k_z, and $g^-(k)$ for negative. Apply diffusive scattering at the boundaries, e.g. $g(k;z=0,t)=0$, to find the undetermined coefficients.
   c. Write down the z-dependent integral expression for current flow in polar coordinates along the z axis where $g^+(k)$ is evaluated for $\theta=0..\pi/2$ and $g^-(k)$ from $\pi/2..\pi$.
   d. Perform the azimuthal integration and evaluate the thickness-averaged value $\frac{1}{t}\int_0^t j(z)dz$ (still an unresolved integral over the Fermi surface polar angle θ)
   e. The resulting integral cannot be evaluated analytically, except in two limits: $\lambda \gg t$ ("thin" limit) and $\lambda \ll t$ ("thick" limit). Perform these integrations, using the variable substitution $\xi=1/\cos(\theta)$

3. Using Matlab or similar, numerically evaluate the integral for 1D current flow
   $$I(V) = 2e \int_{k_{min}}^{k_F} \frac{\hbar k}{m} T(E(k),V) \frac{dk}{2\pi} \quad \text{through:}$$
   a. a single tunnel barrier, 0.5nm thick.
   b. a double barrier with resonance below the Fermi energy. For example, two 0.1nm-thick barriers separated by 0.25nm.

   Use 2eV Fermi energy and 4eV barrier height. For (a), you can use the WKB tunneling probability. However, for (b) this technique is insufficient because as an incoherent method it does not reproduce the resonance. You may want to use a transfer matrix method; see "General Resources" on our piazza.com page.

1) $j_s = A^* T^2 e^{-\Phi/k_B T}$

$\longrightarrow \ln \frac{j_s}{T^2} = -\Phi/k_B T + \ln A^*$

So plotting $\ln j_s/T^2$ vs $\frac{1}{k_B T}$ gives a line whose slope is given by $-\Phi$. For noisy data, linear regression yields best fit and standard error:

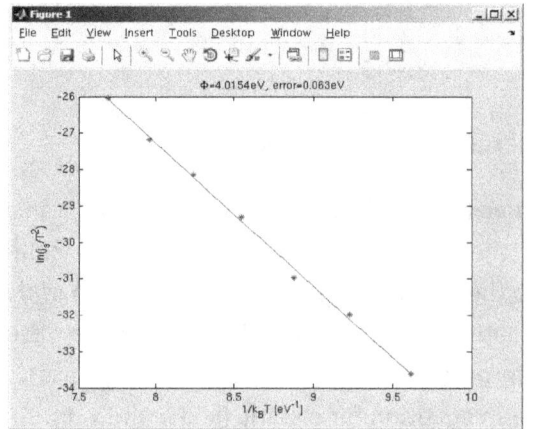

2) see Eqns. 11-18 in Fuch's original paper:
http://piazza.com/umd/fall2013/phys731/resources#   (fuchs.pdf)

Note that for large $K = t/\lambda$ (Thick film) the correction to $\sigma \sim \sigma_0$ in Eqn 18 is dominated by the third term and
$$\sigma \cong \sigma_0 (1 - \tfrac{3}{8} \tfrac{\lambda}{t})$$

For small $K = t/\lambda$ (thin film), leading orders of each term are

$$\frac{\sigma}{\sigma_0} = 1 + \tfrac{3}{4} K B(K) - \tfrac{3}{8 K} \cdot \cancel{K} - \tfrac{5}{8} = \tfrac{3}{4} K B(K)$$

where $B(K) = \int_K^\infty \frac{e^{-\xi}}{\xi} d\xi \sim \int_K^\infty \frac{1 - \xi + \frac{\xi^2}{2} - \cdots}{\xi} d\xi$

$= \ln \xi - \xi + \tfrac{\xi^2}{4} - \tfrac{\xi^3}{18} + \cdots \Big|_K^\infty$

The value at infinity vanishes and we are left with $-\ln K = \ln \tfrac{1}{K}$
so $\sigma \sim \sigma_0 \cdot \tfrac{3}{4} \tfrac{t}{\lambda} \ln \tfrac{\lambda}{t}$

3) See lecture 8, p. 5+6.
For Matlab integration example, see
http://piazza.com/umd/fall2013/phys731/resources#

Boundary Conditions at an arbitrary interface

$$Ae^{ik_1x} \Rightarrow \quad | \quad Ce^{ik_2x} \Rightarrow$$
$$+Be^{-ik_1x} \Leftarrow \quad | \quad +De^{-ik_2x} \Leftarrow$$

$$\left(K_1 = \sqrt{\tfrac{2m(E-V_1)}{\hbar^2}}\right) \quad x \quad \left(K_2 = \sqrt{\tfrac{2m(E-V_2)}{\hbar^2}}\right)$$

ψ continuous: $Ae^{ik_1x} + Be^{-ik_1x}\big|_{bdry} = Ce^{ik_2x} + De^{-ik_2x}\big|_{bdry}$

ψ' continuous: $ik_1 Ae^{ik_1x} - ik_1 Be^{-ik_1x}\big|_{bdry} = ik_2 Ce^{ik_2x} - ik_2 De^{-ik_2x}\big|_{bdry}$

A matrix equation!:

$$\begin{bmatrix} e^{ik_1x} & e^{-ik_1x} \\ ik_1 e^{ik_1x} & -ik_1 e^{-ik_1x} \end{bmatrix} \begin{bmatrix} A \\ B \end{bmatrix} = \begin{bmatrix} e^{ik_2x} & e^{-ik_2x} \\ ik_2 e^{ik_2x} & -ik_2 e^{-ik_2x} \end{bmatrix} \begin{bmatrix} C \\ D \end{bmatrix}$$

$$\begin{bmatrix} A \\ B \end{bmatrix} = \begin{bmatrix} \tfrac{1}{2}e^{-ik_1x} & -\tfrac{1}{2}\tfrac{i}{k_1}e^{-ik_1x} \\ \tfrac{1}{2}e^{ik_1x} & \tfrac{1}{2}\tfrac{i}{k_1}e^{ik_1x} \end{bmatrix} \begin{bmatrix} e^{ik_2x} & e^{-ik_2x} \\ ik_2 e^{ik_2x} & -ik_2 e^{-ik_2x} \end{bmatrix} \begin{bmatrix} C \\ D \end{bmatrix}$$

$$\begin{bmatrix} A \\ B \end{bmatrix} = \begin{bmatrix} \left(\tfrac{1}{2}+\tfrac{k_2}{2k_1}\right)e^{i(k_2-k_1)x} & \left(\tfrac{1}{2}-\tfrac{k_2}{2k_1}\right)e^{-i(k_2+k_1)x} \\ \left(\tfrac{1}{2}-\tfrac{k_2}{2k_1}\right)e^{i(k_2+k_1)x} & \left(\tfrac{1}{2}+\tfrac{k_2}{2k_1}\right)e^{i(k_1-k_2)x} \end{bmatrix} \begin{bmatrix} C \\ D \end{bmatrix}$$

For an arbitrary scattering potential: $V(x)$

$\begin{bmatrix} A \\ B \end{bmatrix}$ ... $\begin{bmatrix} Y \\ Z \end{bmatrix}$

interface: 1 2 ..... N

$$\begin{bmatrix} A \\ B \end{bmatrix} = \begin{bmatrix} \ \end{bmatrix}\begin{bmatrix} \ \end{bmatrix}\cdots\begin{bmatrix} \ \end{bmatrix}\begin{bmatrix} \ \end{bmatrix}\begin{bmatrix} Y \\ Z \end{bmatrix} = \hat{M}\begin{bmatrix} Y \\ Z \end{bmatrix}$$

interface 1, interface 2, interface N-1, interface N

Contains all we need to know about potential!

Transmission Coefficient

$$\begin{bmatrix} A=1 \\ B=r \end{bmatrix} = \begin{bmatrix} M_{11} & M_{12} \\ M_{21} & M_{22} \end{bmatrix} \begin{bmatrix} Y=t \\ Z=0 \end{bmatrix}$$

$$1 = t \cdot M_{11} \implies t = \frac{1}{M_{11}}$$

$$r = t \cdot M_{21} \implies r = \frac{M_{21}}{M_{11}}$$

$$T = \frac{J_{trans}}{J_{inc}} = \frac{\frac{\hbar k_R}{m}|t|^2}{\frac{\hbar k_L}{m}} = \frac{k_R}{k_L}\left|\frac{1}{M_{11}}\right|^2, \quad R = \left|\frac{M_{21}}{M_{11}}\right|^2$$

"Recipe":

For a given piecewise-constant scattering potential:

1. Pick E

2. Construct 2x2 matrix for each interface from E, V_1, V_2, and position x

3. Multiply them together (in proper order) to get $\hat{M}$

4. Calculate $T(E) = \frac{k_R}{k_L}\frac{1}{|M_{11}|^2}$

5. Go to #1 (repeat for different E)

"Double Barrier"

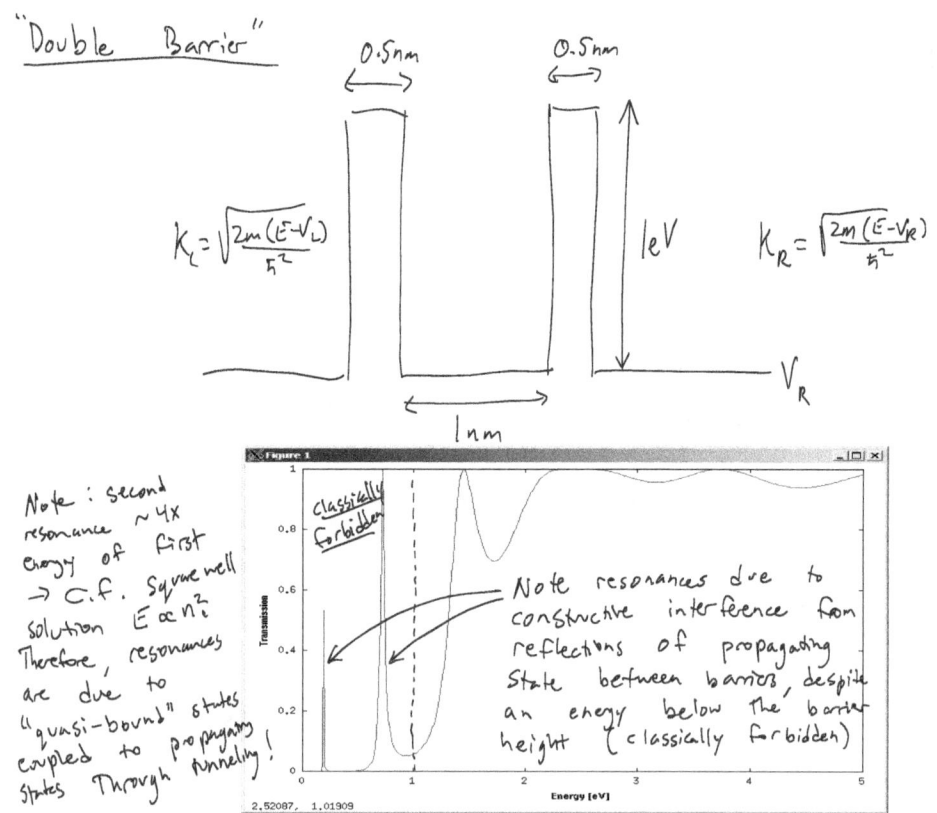

$K_L = \sqrt{\frac{2m(E-V_L)}{\hbar^2}}$ $|eV$ $K_R = \sqrt{\frac{2m(E-V_R)}{\hbar^2}}$

0.5nm, 0.5nm, 1nm, V_R

Note: second resonance ~4× energy of first → c.f. square well solution $E \propto n^2$. Therefore, resonances are due to "quasi-bound" states coupled to propagating states through tunneling!

Note resonances due to constructive interference from reflections of propagating state between barriers, despite an energy below the barrier height (classically forbidden)

Homework 3:

1. Sommerfeld Expansion in 2D
    a. Show that every term in the Sommerfeld expansion for density n vanishes except for the $T=0$ term. Also show that this procedure (erroneously) predicts the exact relationship $\mu=E_F$.
    b. By performing the exact integration analytically, derive the following result:
       $\mu + k_B T \cdot \ln(1+\exp(-\mu/k_B T)) = E_F$
    c. Using a suitable expansion of the natural log, show that the correction from $\mu=E_F$ is negligible.
    d. The conclusion in (a) included all orders of the expansion – so why is there a discrepancy with the exact solution?

2. Free electron gas with an anisotropic dispersion
    a. Suppose one has a free electron gas in 3D where the dispersion relation is given by $E(\vec{k}) = \frac{\hbar^2 k_x^2}{2m_x} + \frac{\hbar^2 k_y^2}{2m_y} + \frac{\hbar^2 k_z^2}{2m_z}$. Note that the electron has a different "mass" associated with its kinetic energy when moving in different directions. This happens in materials as a result of the interaction of the electrons with the atoms. Find the density of states function $D(E)$ so that an integral over k-space can be converted into an integral over energy.
    b. Repeat for 2D gas.

3. deHaas-vanAlphen effect
Determine the constraints on the following to avoid degradation of the oscillatory structure of the dH-vA effect:
    a. Thermal fluctuations ($k_B T$)
    b. Magnetic field inhomogeneities ($\Delta H/H$)

4. Pauli Susceptibility
By Taylor expansion of the density of states, calculate the Pauli susceptibility via evaluation of the integral expressions for spin-up and spin-down electron density.

1. Sommerfeld expansion in 2D

(a)
$$n = \int_0^\infty g(E) f(E) dE \xrightarrow{\text{Sommerfeld}} \int_0^\mu g(E) dE + \sum_{n=1}^\infty \frac{d^{2n-1} g(E)}{dE^{2n-1}} \int \frac{(E-\mu)^{2n}}{(2n)!} \left(-\frac{df}{dE}\right) dE$$

↑ all derivatives of $g_{2D}(E) = \text{const}$ are zero!

Since n is temperature-independent, $n = \int_0^{E_F} g(E) dE = \int_0^\mu g(E) dE$

so $\mu = E_F$, which is incorrect

(b)
$n = \int_0^\infty g(E) f(E) dE$. In 2D, $g(E) = \frac{m}{\pi \hbar^2}$ (see lecture 11, p.2)

Therefore, $n = \frac{m}{\pi \hbar^2} \int_0^\infty \frac{dE}{e^{(E-\mu)/k_B T} + 1} = \frac{m}{\pi \hbar^2} \int_0^\infty \frac{e^{-(E-\mu)/k_B T} dE}{1 + e^{-(E-\mu)/k_B T}}$

$= \frac{m}{\pi \hbar^2} \int_0^\infty \frac{(-k_B T) d(e^{-(E-\mu)/k_B T})}{1 + e^{-(E-\mu)/k_B T}} = -\frac{m k_B T}{\pi \hbar^2} \int_{e^{\mu/k_B T}}^0 \frac{dx}{1+x} = -\frac{m k_B T}{\pi \hbar^2} \ln(1+x) \Big|_{e^{\mu/k_B T}}^0$

$= \frac{m k_B T}{\pi \hbar^2} \ln\left[1 + e^{\mu/k_B T}\right]$

This must be equal to the $T=0$ result $n = \int_0^{E_F} g(E) dE = \frac{m E_F}{\pi \hbar^2}$

Therefore, $E_F = k_B T \ln[1 + e^{\mu/k_B T}] = k_B T \left(\ln(e^{\mu/k_B T}) + \ln(e^{-\mu/k_B T} + 1) \right)$

$= \mu + \ln(1 + e^{-\mu/k_B T})$

(c) For high-density metals, $\mu \gg k_B T$ so $e^{-\mu/k_B T}$ is small. Since $\ln(1+x) \sim x - \frac{x^2}{2} + \ldots$, our correction to $\mu = E_F$ is $e^{-\mu/k_B T} \sim e^{-100} \to$ quite negligible!

(d) The logarithmic deviation increases slower than any power law and so is not captured by the power-law Taylor expansion.

2. Anisotropic dispersion

(a) Define $q_x = \sqrt{\frac{m}{m_x}} k_x$, $q_y = \sqrt{\frac{m}{m_y}} k_y$, $q_z = \sqrt{\frac{m}{m_z}} k_z$ so that

$$E = \frac{\hbar^2}{2m}(q_x^2 + q_y^2 + q_z^2) = \frac{\hbar^2}{2m}|q|^2.$$

Then $\frac{dk_x dk_y dk_z}{(2\pi)^3} = \frac{\sqrt{m_x m_y m_z}}{m^{3/2}} \frac{dq_x dq_y dq_z}{(2\pi)^3}$ so the density of states is

$$\frac{\sqrt{m_x m_y m_z}}{m^{3/2}} \cdot g_{3D}(E) = \frac{\sqrt{m_x m_y m_z}}{m^{3/2}} \cdot \frac{1}{2\pi^2} \left(\frac{2m}{\hbar^2}\right)^{3/2} E^{1/2}$$

So same as in isotropic case, except

$m \to (m_x m_y m_z)^{1/3}$ ("density of states effective mass")

(b) In 2D, the same calculation yields $m \to \sqrt{m_x m_y}$

3. deHaas–van Alphen effect

(a) Thermal fluctuations: Landau level spacing $\frac{\hbar eB}{m} > k_B T$

Therefore we don't expect to see dHvA oscillations at RT unless $B \gtrsim \frac{k_B T}{2\mu_B} = \frac{1\,eV}{2 \cdot 40 \cdot 5.8\times 10^{-5} eV/T} \sim 200\,\text{Tesla}!$ (not obtainable in lab)

At low temp $\sim 1.5K = \frac{RT}{200}$, oscillations will be visible @ a more-reasonable 1T.

(b) Field inhomogeneities: Even at $T=0$, these will smear out oscillations. We must have $\Delta H <$ oscillation period $= \frac{2\pi e}{S_F \hbar} H^2$

Therefore $\frac{\Delta H}{H} < \frac{2\pi e}{S_F \hbar} H$.

4. Pauli Susceptibility

$$M = \left[\frac{1}{2}\int_{-\mu_B H}^{E_F} D(E+\mu_B H)\,dE - \frac{1}{2}\int_{\mu_B H}^{E_F} D(E-\mu_B H)\,dE \right]\mu_B$$

Expand $D(E \pm \mu_B H) \sim D(E) \pm \mu_B H\, D'(E)$

Then
$$M = \mu_B\left[\frac{1}{2}\underbrace{\int_{-\mu_B H}^{E_F} D(E)\,dE - \frac{1}{2}\int_{\mu_B H}^{E_F} D(E)\,dE}_{\sim 0 \text{ if } \mu_B H \text{ small}} + \frac{1}{2}\int_{-\mu_B H}^{E_F}\mu_B H\,D'(E)\,dE + \frac{1}{2}\int_{\mu_B H}^{E_F}\mu_B H\,D'(E)\,dE \right]$$

Since $\int_a^b D'(E)\,dE = D(E)\Big|_a^b$,

$$= \mu_B\left[\frac{1}{2}\mu_B H\,(D(E_F) - \cancel{D(-\mu_B H)}^0) + \frac{1}{2}\mu_B H\,(D(E_F) - \cancel{D(\mu_B H)}^{small})\right]$$

$$= \mu_B^2 H\,D(E_F) = \left(\frac{3}{2}\mu_B^2 \frac{n}{E_F}\right) H$$

Homework 4:

1. Rather than starting with a semiclassical approximation, a quantum mechanical approach to screening in the electron gas would first solve Schroedinger's equation and then determine the induced charge density from the resulting wavefunction state occupation.

    a. Write down the wavefunctions resulting from a perturbing potential V(r) using first-order time-independent perturbation theory. Use your knowledge of the unperturbed wavefunctions to simplify. Your answer will be in terms of an undetermined sum over states.

    b. Using this first-order wavefunction and the Fermi-Dirac occupation function f(k), determine the real-space charge density $\delta\rho(r)$ to first order in V(r) by summing again over all states. Do not attempt to evaluate either sum (yet). Subtract the unperturbed background.

    c. Now find the Fourier transform $\widetilde{\delta\rho}(q) = \int \delta\rho(r) e^{-iq\cdot r} d^3r$. Note that $\int e^{iq'r} e^{-iqr} dr = \delta(q-q')$, which enables the evaluation of the sum over one state index.

    d. Expand the denominator of your expression to obtain $\widetilde{\delta\rho}(q) = -2e\tilde{V}(q) \int \frac{d^3k}{(2\pi)^3} \frac{f\left(k-\frac{q}{2}\right) - f\left(k+\frac{q}{2}\right)}{\frac{\hbar^2}{m}k\cdot q}$. This is the "Lindhard" screening charge density.

    e. Expand the numerator to first order in q and perform the integral in the zero-temperature limit to recover the Thomas-Fermi result $\delta\rho(r) = e^2 g(E_F) \delta\phi(r)$.

    f. The Thomas-Fermi screening charge density is obviously only leading order to the more correct Lindhard result. Without direct calculation, comment on the corrections you expect to $\delta\rho(r)$ from exact integration over the Fermi sphere.

2. Antiferromagnetism is the result of a **negative** exchange constant J<0 causing alternating up/down magnetic moments of nearest neighbors on a lattice.

    a. By modeling this system as two interpenetrating lattices "A" and "B", subject to different mean fields (parameterized through symmetric exchange constants w) coupling them through nearest-neighbor interactions, write the appropriate expressions determining the constituent magnetizations M_A and M_B.

    b. Examine the high-temperature/ low field behavior of M_A and M_B by a suitable approximation.

    c. Solve the system of two linear equations found above and find the total magnetic susceptibility of the antiferromagnet. Identify quantities having units of temperature and comment on the temperature behavior of the inverse susceptibility relative to paramagnetic and ferromagnetic systems.

3. We used a simple two-state spin-1/2 model in our calculation of paramagnetic susceptibility for localized magnetic moments. What if the spin quantum number was so large (>> ½) that we are in the classical limit where orientation of magnetic moment varies continuously instead of discretely?

a. Write an integral expression for the magnetization M=N<μ_z> in a magnetic field H using classical statistical mechanics.
b. Evaluate the integrals to find M=Nμ$L(\alpha)$, where $\alpha = \mu H/k_B T$, and $L(\alpha)$ is the "Langevin function"
c. Recover Curie's Law from the low field/ high temperature limit of M.
d. By noting that $L(\alpha)$ is the limit of a sequence of a type of special function, infer the susceptibility of a quantum system of arbitrary angular momentum >1/2.

4. A mean-field theory approach to spin lattices ordered by local exchange ignores local fluctuations near the critical temperature. An exact approach to this so-called Ising model shows that these fluctuations eliminate a phase transition at nonzero temperature in 1-dimension.
   a. Construct a numerical model of the 2-dimensional Ising system, with at least 25x25 2-level spins and periodic boundary conditions. With a constant and isotropic exchange constant J, implement the Metropolis algorithm:
      i. Initialize a perfectly ordered spin lattice.
      ii. Choose a spin at random, and evaluate the change in energy of the system, ΔE, when the spin is flipped.
      iii. If $\Delta E<0$, then flip the spin. Otherwise, flip the spin with probability $P=exp(-\beta \Delta E)$.
      iv. Repeat until "thermal equilibrium" is achieved.
   b. Plot the magnetization, net energy, and heat capacity from the thermalized distribution as a function of temperature.
   c. Compare the "critical temperature" obtained numerically with the prediction of mean-field theory and the *exact* result $\frac{k_B T_c}{J} = \frac{2}{\ln(1+\sqrt{2})}$ derived by Onsager in 1944.
   d. Likewise, compare the magnetization for T<T_c, with the exact result $M=[1-sinh^{-4}(2\beta J)]^{1/8}$

5. Ferromagnetic hysteresis
   a. Numerically solve the energy minimization of a uniaxial magnetic domain for arbitrary orientation θ with respect to an applied magnetic field. Plot out the M-H relations for at least 5 values of θ on the same figure.
   b. Using your code from above, calculate the mean coercive field and the mean remnant field from an ensemble of randomly oriented domains.

1) A quantum mechanical approach to screening in the electron gas would first solve Schroedinger's eqn and determine the induced charge density from the perturbed wavefunctions.

$$-\frac{\hbar^2}{2m}\nabla^2\psi + V\psi = E\psi \xrightarrow{\text{find }\psi} \delta\rho(r) \xrightarrow[\text{transform}]{\text{Fourier}} \delta\tilde{\rho}(q) \longrightarrow \epsilon(q) = 1 + \frac{\widetilde{\delta\rho(q)}}{\epsilon_0 \delta\tilde{\phi}(q)}$$

ⓐ Write down the wavefunction resulting from a perturbing potential $V(\vec{r})$ using first-order time-independent perturbation Thy. Use your knowledge of the unperturbed wavefunctions to simplify.

$$\psi(k) = \psi^0(k) + \sum_{k'} \frac{\int \psi^{0*}(k') V(r) \psi^0(k) d^3r}{E_k - E_{k'}} \psi^0(k')$$

Since $\psi^0(k) = e^{ikr}$,

$$\psi(k) = e^{ikr} + \sum_{k'} \frac{\int V(r) e^{i(k-k')r} d^3r}{E_k - E_{k'}} e^{ik'r}$$

$$= e^{ikr} + \sum_{k'} \frac{\tilde{V}(k-k')}{E_k - E_{k'}} e^{ik'r}$$

ⓑ Using this first-order wavefunction and the Fermi-Dirac occupation function $f(k)$, determine the charge density $\rho(r)$ to first order in V. Do not attempt to evaluate sums over states (yet). Subtract unperturbed background.

$$\delta\rho(r) = 2e\left[\sum_k f(k)|\psi(k)|^2 - \sum_k f(k)|\psi^0(k)|^2\right] = 2e \sum_k f(k) \left[\sum_{k'} \frac{\tilde{V}(k'-k)}{E_k - E_{k'}} e^{i(k-k')r} + \sum_{k'} \frac{\tilde{V}(k-k')}{E_k - E_{k'}} e^{i(k'-k)r}\right]$$

(c) Now find the Fourier transform $\tilde{\rho}(q) = \int \rho(\vec{r}) e^{-iq\cdot r} d^3r$. Note that $\int (e^{iq'r}) e^{-iqr} d^3r = \delta(q-q')$ which enables the evaluation of the sum over one state index.

$$\delta\tilde{\rho}(q) = 2e \sum_k f(k) \left[\sum_{k'} \frac{\tilde{V}(k'-k)}{E_k - E_{k'}} \delta(q-(k'-k)) + \sum_{k'} \frac{\tilde{V}(k-k')}{E_k - E_{k'}} \delta(q-(k-k')) \right]$$

$$\delta\tilde{\rho}(q) = 2e \sum_k f(k) \left[\frac{\tilde{V}(q)}{E_k - E_{k+q}} + \frac{\tilde{V}(q)}{E_k - E_{k-q}} \right]$$

$$= 2e\tilde{V}(q) \left(\int \frac{d^3k}{(2\pi)^3} \frac{f(k-q/2)}{E_{k-q/2} - E_{k+q/2}} + \int \frac{d^3k}{(2\pi)^3} \frac{f(k+q/2)}{E_{k+q/2} - E_{k-q/2}} \right)$$

$$= 2e\tilde{V}(q) \int \frac{d^3k}{(2\pi)^3} \frac{f(k-q/2) - f(k+q/2)}{E_{k-q/2} - E_{k+q/2}}$$

(d) Expand the denominator to obtain

$$\delta\tilde{\rho}(q) \cong 2e\tilde{V}(q) \int \frac{d^3k}{(2\pi)^3} \frac{f(k-q/2) - f(k+q/2)}{\frac{\hbar^2}{2m}\left[(k^2 - k\cdot q + \frac{q^2}{4}) - (k^2 + k\cdot q + \frac{q^2}{4})\right]}$$

$$= -2e\tilde{V}(q) \int \frac{d^3k}{(2\pi)^3} \frac{f(k-q/2) - f(k+q/2)}{\frac{\hbar^2}{m} k\cdot q}$$

This is the "Lindhard" screening charge density.

(e) Expand the numerator to first order in q and perform the integral in the zero-temperature limit to recover our Thomas-Fermi result.

$$\delta\tilde{\rho}(q) \cong -2e\tilde{V}(q) \int \frac{d^3k}{(2\pi)^3} \frac{\left[\left(f(k) - \frac{df}{dE}\frac{dE}{dk}\frac{q}{2}\right) - \left(f(k) + \frac{df}{dE}\frac{dE}{dk}\frac{q}{2}\right)\right]}{\frac{\hbar^2}{m} k\cdot q}$$

$$= -e\tilde{V}(q) \int \left(\overset{D(E)dE}{\overbrace{\frac{2 d^3k}{(2\pi)^3}}}\right) \frac{\frac{\hbar^2}{m} k\cdot q \frac{df}{dE}}{\frac{\hbar^2}{m} k\cdot q} \rightarrow -e\delta\tilde{\phi}(q) \int D(E) \frac{df}{dE} dE$$

$$= e^2 D(E_F) \delta\tilde{\phi}(q) \qquad \text{So,} \qquad \epsilon(q) = 1 + \frac{\delta\tilde{\rho}(q)}{\epsilon_0 \tilde{\phi}} = 1 + \frac{e^2 D(E_F)}{\epsilon_0}$$

(f) This result is only leading order. Comment on the corrections you expect to $\delta\rho(r)$ from exact integration over the Fermi sphere in the $T=0$ limit.

Integration over sharp Fermi surface gives "Friedel oscillations" similar to oscillations in exchange hole density.

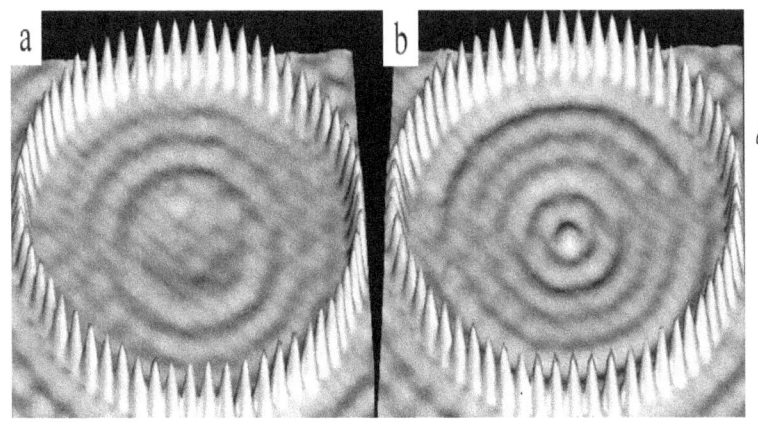

STM of "quantum corral" (Fe on Cu)

2.) Antiferromagnetism is the result of a negative exchange constant causing alternating up/down magnetic moments on a lattice.

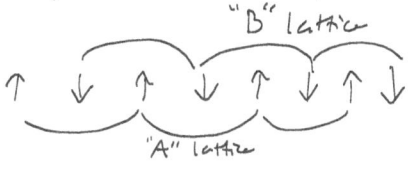

(a) By modeling this system as two interpenetrating lattices "A" and "B", subject to two different mean fields due to the nearest-neighbor exchange interaction, write the appropriate expressions determining M_A and M_B.

Mean fields: $H_A = -wM_B$ and $H_B = -wM_A$

So $M_A = \frac{N}{2}\mu \tanh\left(\frac{\mu(H+H_A)}{k_B T}\right)$ $\quad M_B = \frac{N}{2}\mu \tanh\left(\frac{\mu(H+H_B)}{k_B T}\right)$

(b) Examine the high-temperature behavior of M_A and M_B by a suitable approximation

to first order,
$$M_A \sim \frac{N\mu^2(H-wM_B)}{2k_BT} \qquad M_B \sim \frac{N\mu^2}{2k_BT}(H-wM_A)$$

(c) Solve the system of two linear equations found above and find the magnetic susceptibility. Identify quantities having units of temperature.

$$M_A = \frac{\alpha - \beta M_B}{T} = \frac{\alpha - \beta\left(\frac{\alpha - \beta M_A}{T}\right)}{T} = \frac{T\alpha - \beta\alpha + \beta^2 M_A}{T^2}$$

$$M_A = \frac{T\alpha - \beta\alpha}{T^2 - \beta^2} = \frac{\alpha(T-\beta)}{(T-\beta)(T+\beta)} = \frac{\alpha}{T+\beta} \qquad \alpha \equiv \frac{N\mu^2 H}{2k_B}, \; \beta \equiv \frac{N\mu^2 w}{2k_B}$$

So $\quad M = M_A + M_B = \chi H = \frac{C}{T+\beta} H \qquad$ where $\quad C = \frac{N\mu^2}{k_B}$

This is same as Ferromagnetism except w/ a <u>negative</u> Curie-Weiss critical temperature. In $H=0$, note that $M_A \sim -\frac{\beta}{T}M_B$ and $M_B = -\frac{\beta}{T}M_A$. The only way to solve this system is $M_A = M_B = 0$ unless $T = \beta$, where $M_A = -M_B \neq 0$. Below $T = \beta = T_N$ (the Néel temperature) the sublattices are magnetically ordered!

3.)

We used a simple two-state spin-1/2 model for our calculation of paramagnetic susceptibility for localized magnetic moments. What if the total angular momentum is so large that we are in the classical limit where orientation of magnetic moment $\vec{\mu}$ varies continuously?

(a) Write an (integral) expression for the magnetization $M = N\langle \mu_z \rangle$ in a magnetic field using classical stat. mech.

$$M = N\mu \langle \cos\theta \rangle = N\mu \frac{\int_0^\pi \cos\theta \cdot e^{\frac{\mu H \cos\theta}{k_B T}} \cdot \sin\theta \, d\theta}{\int_0^\pi e^{\frac{\mu H \cos\theta}{k_B T}} \cdot \sin\theta \, d\theta}$$

(b) Evaluate the integrals to find $M = N\mu L(\alpha)$ where $\alpha = \frac{\mu H}{k_B T}$ and $L(\alpha)$ is the "Langevin function".

Using $x = \cos\theta$,

$$\frac{M}{N\mu} = \frac{\int_{-1}^{1} e^{\alpha x} x \, dx}{\int_{-1}^{1} e^{\alpha x} dx} = \frac{\frac{d}{d\alpha}\int_{-1}^{1} e^{\alpha x} dx}{\int_{-1}^{1} e^{\alpha x} dx} = \frac{d}{d\alpha} \ln\left(\int_{-1}^{1} e^{\alpha x} dx\right)$$

$$\frac{M}{N\mu} = \frac{d}{d\alpha} \ln \frac{e^\alpha - e^{-\alpha}}{\alpha} = \frac{d}{d\alpha} \ln \frac{2\sinh\alpha}{\alpha} = \frac{\alpha}{2\sinh\alpha}\left(\frac{2}{\alpha}\cosh\alpha - \frac{2\sinh\alpha}{\alpha^2}\right)$$

$$= \left(\frac{\cosh\alpha}{\sinh\alpha} - \frac{1}{\alpha}\right) = \coth\alpha - \frac{1}{\alpha} \qquad \text{so } M = N\mu L(\alpha).$$

(c) Recover Curie's law from the low field/high T limit.

$$L(\alpha) \sim \frac{\alpha}{3} - \frac{\alpha^3}{45} + \cdots \qquad \text{so } M = \frac{N\mu^2}{3k_B T} H$$

and $\chi = \frac{C}{T}$

(d) By noting that $L(\alpha)$ is the limit of a sequence of a type of special function, infer the susceptibility of a quantum system of arbitrary angular momentum $> \frac{1}{2}$.

$$M = N\mu B_J(\alpha) \quad \text{where} \quad B_J(\alpha) \text{ is "Brillouin function"}$$

4) Ising model

Metropolis algorithm in Matlab:

```
function [M, E]=ising(N,Jex,T)

kB=0.026/300; %eV/K
kBT=kB*T;

Spin=ones(N); %initialize into ferromagnetically ordered lattice:
              % 1 = 'spin up', -1 = 'spin down'
M=0;Et=0;
for ii=1:(4*N*N) %should equilibrate in ~N*N steps
  randarray=rand(N*N,1); %for selection of random sites
  E=0; %total energy

  for jj=1:(N*N)
    lloc=find(max(randarray)==randarray); %linear indexing, find max
    y=mod(lloc-1,N)+1; %y location
    x=floor(lloc/N-1e-9)+1; %x location

    %nearest neighbor exchange, periodic BCs:
    Ea=Spin(x,mod(y,N)+1); %'East'
    We=Spin(x,mod(y-2,N)+1); %'West'
    No=Spin(mod(x,N)+1,y); %'North'
    So=Spin(mod(x-2,N)+1,y); %'South'

    dE=-Jex*Spin(x,y)*(Ea+We+No+So); %local exchange energy
    if dE>0
      Spin(x,y)=-Spin(x,y); %flip spin
      E=E-dE;
    elseif dE<=0
      if rand<exp(2*dE/kBT) %random flip to higher energy state
        Spin(x,y)=-Spin(x,y);
        E=E-dE;
      else
        E=E+dE;
      end
    end

    randarray(lloc)=0; %remove max valued site so next time finds another
  end

  %Average over many steps after system has reached equilibrium
  if ii>(2*N*N)
    M=M+abs(sum(sum(Spin)));
    Et=Et+E;
  end

end

M=M/(2*N*N)/(N*N); %normalized magnetization
Et=Et/(2*N*N); %total energy

end
```

Results for 25×25 lattice

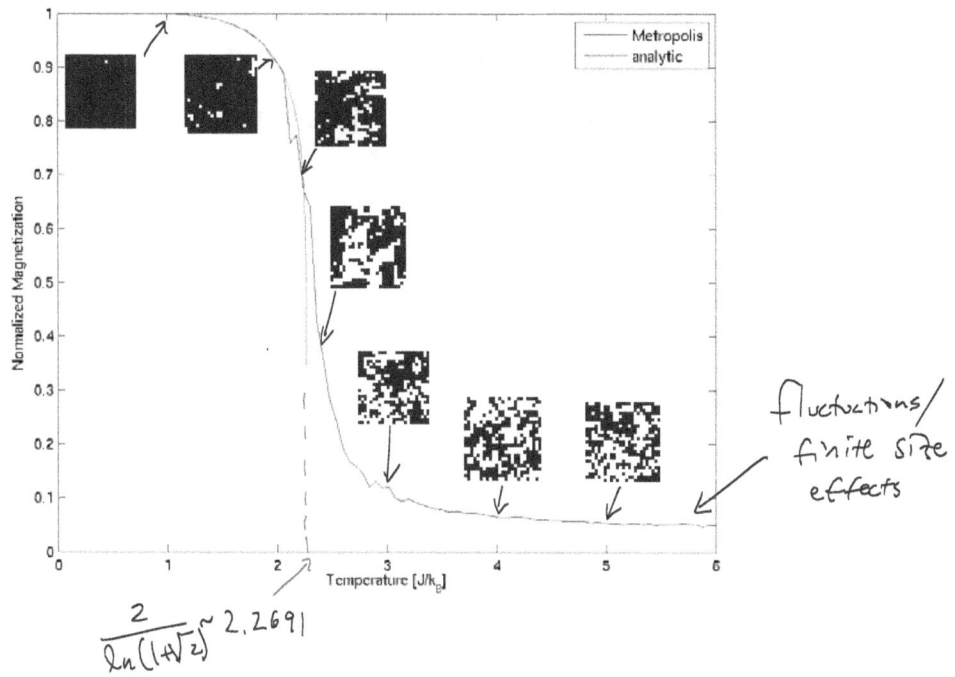

$\frac{2}{\ln(1+\sqrt{2})} \sim 2.2691$

fluctuations/ finite size effects

Energy and Heat Capacity

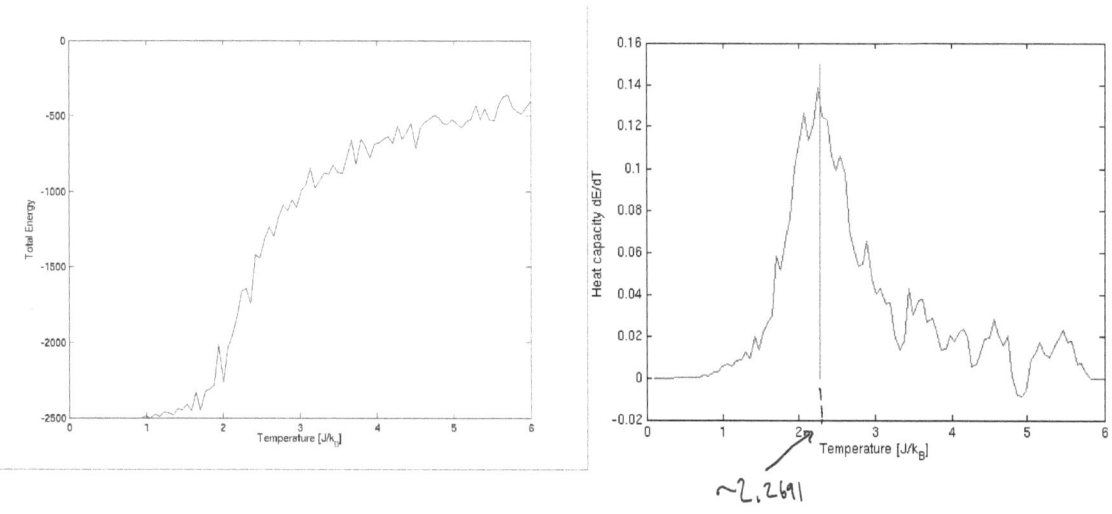

~ 2.2691

5)

```
clear

Nt=10;
Nh=400;
thetas=linspace(0,pi/2,Nt);
hs=linspace(1,-1,Nh);

tt=1;
for theta=thetas
  hh=1;
  for h=hs
    %find root of:
    phi1=fzero(@(phi) 0.5*sin(2*(phi-theta))+h*sin(phi), theta);

    %is it a stable extremum?
    if ((cos(phi1-theta)+h*cos(phi1))>0)
      %did it switch?
      if (hh>1 && cos(phi1)>M(hh-1,tt))
        break;
      else
        H(hh,tt)=h;
        M(hh,tt)=cos(phi1);
        hh=hh+1;
      end
    end
  end

  tt=tt+1;
end

figure(1);
plot(H,M,'.'); hold on;
plot(-H,-M,'.');hold off;
axis([-1.5 1.5 -1.1 1.1])
xlabel('Reduced Magnetic Field h')
ylabel('Magnetization M\cdot cos\phi')
legend(int2str(180/pi*thetas'));

figure(2)
Mtot=zeros(1,length(hs));
for ii=1:(length(hs)-1)
  for jj=1:length(thetas)
    if H(ii,jj)~=0
      Mtot(ii)=Mtot(ii)+M(ii,jj);
    else %it switched
      Mtot(ii)=Mtot(ii)-M(length(hs)-ii,jj);
    end
  end
end
plot(hs,Mtot);hold on;
plot(-hs, -Mtot); hold off;
xlabel('Reduced Magnetic Field h')
ylabel('Magnetization M\cdot cos\phi')

%remanant magnetization: ~0.6360
```

124

Homework 5:

1. Using a matrix eigenvalue/eigenvector solver such as Matlab, construct the Hamiltonian in Fourier representation for the potential $V(x)=0.5*\cos(2\pi x/a)$:
   a. Plot at least 5 bands of the 1-dimensional bandstructure, showing $2\pi/a$ periodicity and inversion symmetry in k-space.
   b. Plot the Brillouin-zone-edge wavefunction probability densities in real-space and compare to the potential.

2. Calculate the "nearly-free electron" bandstructure of :
   a. the 2D square lattice
   b. the 3D FCC lattice. Compare this to the bandstructure of Aluminum.

1) 1D diagonalization of finite periodic matrix hamiltonian

```
%CONSTANTS
c=2.998e10;%cm/s
hbar=6.582e-16; %in eV*sec
m=5.11e5/c^2; %in eV/c^2
hbar2o2m=hbar^2/(2*m);

%UNIT CELL AND LATTICE SPACING
a=15e-8;
dx=0.3e-8;
x=0:dx:(a-dx);
N=round(a/dx);

%POTENTIAL
Vx=0.5*cos(2*pi*x/a);

%POTENTIAL PART OF HAMILTONIAN
R=fft(eye(N));
Hpot=R'*diag(Vx)*R/N;

%%OR
%VG=fft(Vx)/N;
%for ii=1:NXPTS
% Hpot(ii,:)=shift(VG,ii-1);
%end
%OR
%Hpot=zeros(NXPTS);
%for ii=1:(NXPTS-1)
% Hpot=Hpot+diag(VG(NXPTS+1-ii)*ones(NXPTS-ii,1),ii);
%end
%for ii=-(NXPTS-1):0

%LABEL THE RECIPROCAL LATTICE VECTORS
G=pi/a*[-fliplr(2:2:(N-1)) 0:2:N];

%PICK SOME POINTS IN THE FIRST B-ZONE
NKPTS=51;
ZONES=1;
ks=ZONES*linspace(-pi/a,pi/a ,NKPTS*ZONES);

jj=1;
for k = ks

%ADD KINETIC ENERGY TO HAMILTONIAN
H=diag(hbar2o2m*(k+G).^2)+Hpot;

%FIND ENERGY EIGENVALUES
[v,d]=eig(H);
eig1=diag(d);
evals(:,jj)=sort(real(eig1));

jj=jj+1;

end

figure(1)
NUMBANDS=5;
plot(ks*a/pi, evals(1:NUMBANDS,:));

xlabel('k [\pi/a]'); ylabel('Energy [eV]');
axis([-ZONES ZONES -.5 4])

figure(2)
[e1,ei1]=min(real(eig1));
Psi1=conj(ifft(v(:,ei1))).*ifft(v(:,ei1))*N;
plot(x/a,Psi1,'b');%,['b;1st band E=' num2str(real(e1)) ';']);
hold on;
eig2=eig1; eig2(ei1)=realmax;
[e2,ei2]=min(eig2);
Psi2=conj(ifft(v(:,ei2))).*ifft(v(:,ei2))*N;
plot(x/a,Psi2,'r');%,['r;2nd band E=' num2str(real(e2)) ';'])
plot(x/a,Vx/max(Vx)/N,'k');%;k;V(x);');
title('zone-edge realspace electron density')
hold off;
xlabel('Distance [a]');
ylabel('\Psi^* \Psi')
legend('1st band','2nd band','V(x)')
```

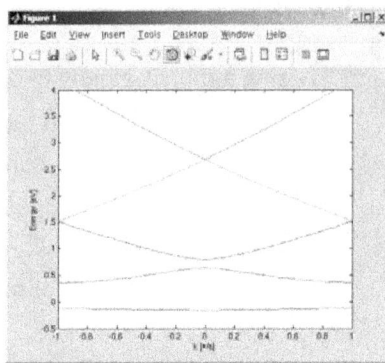

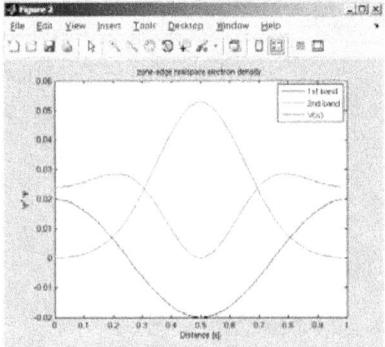

2a) nearly free electrons in 2D square lattice

```
N=2;
kk=1;
for ii=-N:N
 for jj=-N:N
  kps(kk,:)=[ii jj];
  kk=kk+1;
 end
end

N=25;
Emax=3;

ks= [[zeros(N,1) linspace(0,0.5,N)']; ...
     [linspace(0,0.5,N)' 0.5*ones(N,1)]; ...
     [linspace(0.5,0,round(sqrt(2)*N))' linspace(0.5,0,round(sqrt(2)*N))']];

for ii=1:length(ks)
 for jj=1:length(kps)
  E(ii,jj)=norm(ks(ii,:)+kps(jj,:))^2;
 end
end

plot(E,'k')
ylabel('Energy  [h^2/(2ma^2)]')
axis([0 length(ks) 0 Emax])
hold on;
text(0, -0.1,'\Gamma')
plot([N N], [0 Emax]);
text(N,-0.1, 'X')
plot(2*[N N], [0 Emax])
text(2*N,-0.1, 'M')
text((2+sqrt(2))*N, -0.1,'\Gamma')
set(gca,'xtick',[])
set(gca,'xticklabel',[])
```

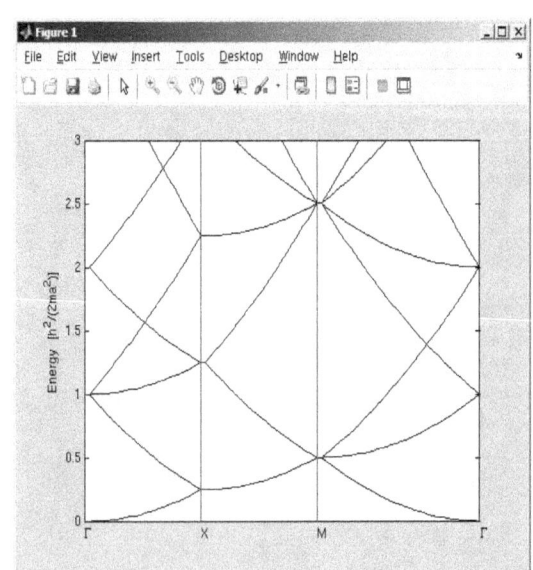

2b) Nearly-free electrons in 3D: FCC lattice

```
%In units of 2*pi/a: basis vectors of FCC reciprocal space (BCC) in cubic lattice
d1=[0 0 0]; d2=[0.5 0.5 0.5];
%generate list of reciprocal lattice points (Gs)
N=1;Gs=[];
for h=-N:N
  for k=-N:N
    for l=-N:N
      D1=d1+[h k l]; %using conventional cell tiling
      D2=d2+[h k l];
      Gs=[Gs; D1; D2];
    end
  end
end
%plot out unit cell of reciprocal lattice by selecting those close to origin
figure(1); MAXRAD=1;
unitcell=find(sqrt(sum(abs(Gs).^2,2))<MAXRAD);
plot3(Gs(unitcell,1), Gs(unitcell,2), Gs(unitcell,3),'o'); hold on;

N=300; %k points along path
Emax=1;

%High symmetry points in BZ
L=0.25*[1 1 1];
Gamma=[0 0 0];
X=0.5*[1 0 0];
U=0.5*[1 1/4 1/4];
K=[3/8 0 3/8];
W=0.5*[1 1/2 0];

%choose path through k-space along IBZ edges here
%kpath=[L; Gamma; X; U; K; Gamma];
%kpathstring=char ('L', 'Gamma', 'X', 'U', 'K', 'Gamma');
kpath= [L; Gamma; X; W; Gamma; U; X];
kpathstring= char('L', 'Gamma', 'X', 'W', 'Gamma', 'U', 'X');

ks=[];
for ii=1:(length(kpath)-1)
  thispath=[];
  for jj=1:3
    thispath=[ thispath [linspace(kpath(ii,jj),kpath(ii+1,jj),round(N*norm(kpath(ii,:)-kpath(ii+1,:))))']];
  end
  ks=[ks; thispath];
  text(kpath(ii,1),kpath(ii,2),kpath(ii,3),kpathstring(ii,:));
end
plot3(ks(:,1),ks(:,2),ks(:,3)); hold off; %plot path in unit cell

% calculate energies from parabolic dispersion at each reciprocal
% lattice point
for ii=1:length(ks)
  for jj=1:length(Gs)
    E(ii,jj)=norm(ks(ii,:)+Gs(jj,:))^2;
  end
end

figure(2)
plot(E,'b'); hold on;
set(gca,'XTick',[]) %no x tick marks
ylabel('Energy  [h^2/(2ma^2)]')
axis([1 length(ks) 0 Emax]);

%superimpose vertical lines and labels at symmetry points
pathlength=0;
text(0,-0.05,kpathstring(1));
for ii=2:length(kpath)
  pathlength(ii)=round(N*norm(kpath(ii,:)-kpath(ii-1,:)));
  text(sum(pathlength), -0.05, kpathstring(ii));
end
Vlinesx=repmat(cumsum(pathlength),2,1);
Vlinesy=repmat([0; Emax], 1, length(kpath));
plot(Vlinesx,Vlinesy,'k'); hold off;
```

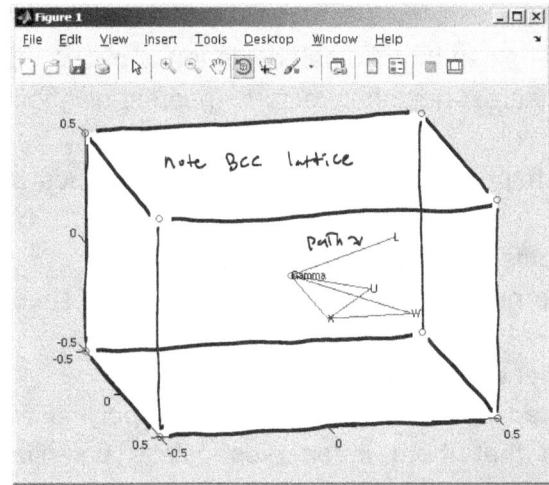

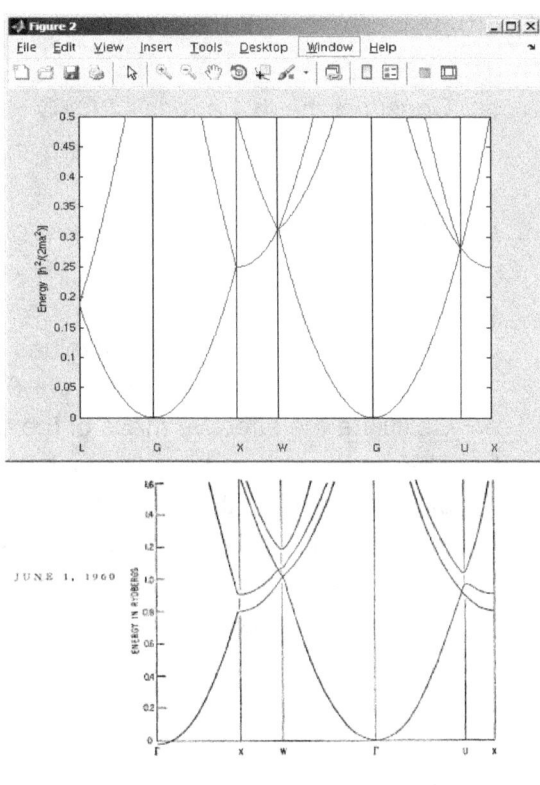

Homework 6:

1. Graphene:
    a. With the p_z basis and 2x2 matrix Hamiltonian used in class, calculate and plot the bandstructure along Γ-K-M-Γ. Also plot the expectation values of σ_x and σ_y for the conduction band wavefunction in the first BZ.
    b. Extend the Hamiltonian to include the s, p_x and p_y orbitals on both A and B sublattice sites, leading to an 8x8 matrix. Because the nearest-neighbor vectors are not orthogonal, you must remember to trigonometrically decompose the overlap matrix elements into σ and π bonds! Recalculate the bandstructure. [One set of overlap parameters is $V_{ss\sigma}$ = -4.80, $V_{sp\sigma}$ = 4.75, $V_{pp\sigma}$ = 4.39, and $V_{pp\pi}$ = -2.56 eV (Phys. Rev. B **70**, 115407 (2004)). Also useful: Phys. Rev. B **74**, 165310 (2006)]
    c. Identify the s,p or sp^2 symmetries of the band wavefunctions at high symmetry points.

2. **Bilayer** graphene consists of two 2-dimensional graphene planes stacked together in the so-called "Bernal" ordering, where an atom in the A lattice of one plane aligns with a B atom in the other:

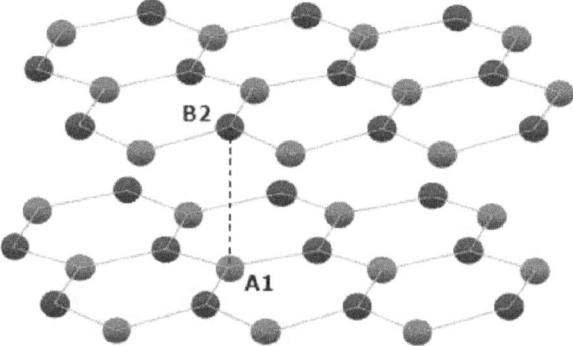

   a. Construct the 4x4 p_z-basis LCAO Hamiltonian for this material in terms of overlap energies $V_{pp\pi}$, $V_{pp\sigma}$ and the in-plane nearest-neighbor vectors. (nearest-neighbor interactions only).
   b. Show that, just like single-sheet graphene, there are two degenerate bands at the K point.
   c. Write out the Hamiltonian matrix elements for states near K in terms of $\Delta \mathbf{k}$.
   d. Show that the energy eigenvalues are not linear as in the case of single-sheet graphene, but parabolic. [hint: $(1+x)^{1/2} \sim 1+x/2-x^2/8$]
   e. Calculate the effective mass of the degenerate bands at the K point.
   f. The two layers in bilayer graphene can be made inequivalent by applying an electric field in the z-direction such that there is an electrostatic potential difference ϕ. What terms in the Hamiltonian will be affected by nonzero ϕ, and what quantity will be added to them?
   g. Show that this asymmetry breaks the degeneracy at K, with an energy splitting of $e\phi$.

3. Calculate the bandstructure of the diamond lattice (FCC with a 2-atom basis) using the LCAO method with these parameters, corresponding to Si:

$$E_s = 0.00$$
$$E_p = 7.20$$
$$V_{ss} = -8.13$$
$$V_{sp} = 5.88$$
$$V_{xx} = 3.17$$
$$V_{xy} = 7.51 \text{ eV}$$

4. Calculate the bandstructure of the 3D simple cubic lattice with sp^3 LCAO, and include spin-orbit interaction which splits the valence band into a light hole, heavy hole, and split-off band. Use

$$E_p = 0$$
$$E_s = 4$$
$$V_{ss} = 0.1$$
$$V_{sp} = 0.1$$
$$V_{pp\sigma} = 0.5$$
$$V_{pp\pi} = 0.3 \text{ eV}$$
spin-orbit energy 0.5 eV

1a)

p_z basis set:

```
clear;

Vpp=-3; %p_z overlap parameter

%nearest-neighbor vectors
n1=[1/sqrt(3) 0];
n2=[-1/(2*sqrt(3)) 1/2];
n3=[-1/(2*sqrt(3)) -1/2];

% basis vectors
d1=[0 0];
d2=n1;

%define path along IBZ perimeter
Gamma=[0 0];
K=[0 4*pi/3];
M=[pi/sqrt(3) pi];
kpath=[Gamma; K; M; Gamma];
kpathstring=char('\Gamma', 'K', 'M', '\Gamma');

N=100; %number of k points from Gamma to K
ks=[];
for ii=1:(length(kpath)-1)
    thispath=[];
    for jj=1:2
        thispath=[ thispath [linspace(kpath(ii,jj),kpath(ii+1,jj), round(N*norm(kpath(ii,:)-kpath(ii+1,:))))']];
    end
    ks=[ks; thispath];
    text(kpath(ii,1)/pi,kpath(ii,2)/pi,kpathstring(ii,:));
end
```

$$\mathcal{H} = \begin{array}{c} \\ \langle p_{z1}| \\ \langle p_{z2}| \end{array} \begin{array}{c} |p_{z1}\rangle \quad |p_{z2}\rangle \\ \begin{bmatrix} E_p & V_{pp\pi}f \\ V_{pp\pi}f^* & E_p \end{bmatrix} \end{array}$$

$$f = \sum_{i=1,2,3} e^{i\vec{k}\cdot\vec{n}_i}$$

```
for kk=1:length(ks)
    k=ks(kk,:);
    f=exp(i*k*n1')+exp(i*k*n2')+exp(i*k*n3');
    H=[0 Vpp*f; Vpp*f' 0];
    eigH(:,kk)=eig(H);
end

Emax=10; TEXTY=0.75;

figure(1);
plot(1:length(ks),eigH); hold on;
set(gca,'XTick',[]) %no x tick marks
ylabel('Energy [eV]')
axis([1 length(ks) -Emax Emax]);
pathlength=0;
text(0,-(Emax+TEXTY),kpathstring(1));
for ii=2:length(kpath)
    pathlength(ii)=round(N*norm(kpath(ii,:)-kpath(ii-1,:)));
    text(sum(pathlength), -(Emax+TEXTY), kpathstring(ii,:));
end
Vlinesx=repmat(cumsum(pathlength),2,1);
Vlinesy=repmat([-Emax; Emax], 1, length(kpath));
plot(Vlinesx,Vlinesy,'k'); hold off;
```

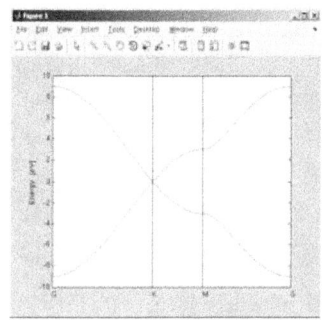

1b) sp^3 basis set

Since on-site orbitals are considered orthogonal, we have:

$$\mathcal{H} = \begin{array}{c} \\ \langle s_1| \\ \langle p_{x1}| \\ \langle p_{y1}| \\ \langle p_{z1}| \\ \langle s_2| \\ \langle p_{x2}| \\ \langle p_{y2}| \\ \langle p_{z2}| \end{array} \begin{array}{c} |s_1\rangle \quad |p_{x1}\rangle \quad |p_{y1}\rangle \quad |p_{z1}\rangle \quad ; \quad |s_2\rangle \quad |p_{x2}\rangle \quad |p_{y2}\rangle \quad |p_{z2}\rangle \\ \left[\begin{array}{cccc|cccc} E_s & 0 & 0 & 0 & & & & \\ 0 & E_p & 0 & 0 & & \mathcal{H}_{12} & & \\ 0 & 0 & E_p & 0 & & & & \\ 0 & 0 & 0 & E_p & & & & \\ \hline & & & & E_s & 0 & 0 & 0 \\ & \mathcal{H}_{21} = \mathcal{H}_{12}^\dagger & & & 0 & E_p & 0 & 0 \\ & & & & 0 & 0 & E_p & 0 \\ & & & & 0 & 0 & 0 & E_p \end{array} \right] \end{array}$$

$\langle s_1 | s_2 \rangle:$ $\quad = V_{ss\sigma}(e^{ikn_1} + e^{ikn_2} + e^{ikn_3})$

$\langle s_1 | p_{x2} \rangle:$ $\quad = V_{sp\sigma} e^{ikn_1} - V_{sp\sigma}\cos 60°(e^{ikn_2} + e^{ikn_3})$

$\langle s_1 | p_{y2} \rangle:$ $\quad = V_{sp\sigma} \cos 30°(e^{ikn_2} - e^{ikn_3})$

$\langle s_1 | p_{z2} \rangle = \langle p_{x1} | p_{z2} \rangle = \langle p_{y1} | p_{z2} \rangle = 0$

$\langle p_{z1} | p_{z2} \rangle = V_{pp\pi}(e^{ikn_1} + e^{ikn_2} + e^{ikn_3})$

$\langle p_{x1} | p_{x2} \rangle:$ $\quad = V_{pp\sigma} e^{ikn_1} + (V_{pp\sigma}\cos^2 60° + V_{pp\pi}\sin^2 60°)(e^{ikn_2} + e^{ikn_3})$

$\langle p_{y1} | p_{y2} \rangle:$ $\quad = V_{pp\pi} e^{ikn_1} + (V_{pp\sigma}\cos^2 30° + V_{pp\pi}\sin^2 30°)(e^{ikn_2} + e^{ikn_3})$

$\langle p_{x1} | p_{y2} \rangle:$ $\quad = \cos 30° \cos 60°(V_{pp\sigma} - V_{pp\pi})(-e^{ikn_2} + e^{ikn_3})$

$\langle p_{y1} | p_{x2} \rangle:$ $\quad = \cos 30° \cos 60°(V_{pp\sigma} - V_{pp\pi})(-e^{ikn_2} + e^{ikn_3})$

So we have:

$$\mathcal{H}_{12} = \begin{array}{c} \\ \langle s_1| \\ \langle p_{x1}| \\ \langle p_{y1}| \\ \langle p_{z1}| \end{array} \begin{pmatrix} |s_2\rangle & |p_{x2}\rangle & |p_{y2}\rangle & |p_{z2}\rangle \\ V_{ss\sigma} f_0 & V_{sp\sigma}(e^{ikn_1} - \tfrac{1}{2}f_1) & \tfrac{\sqrt{3}}{2} V_{sp\sigma} f_2 & 0 \\ -V_{sp\sigma}(e^{ikn_1} - \tfrac{1}{2}f_1) & V_{pp\sigma}(e^{ikn_1} + \tfrac{1}{4}f_1) + \tfrac{3}{4}V_{pp\pi}f_1 & -\tfrac{\sqrt{3}}{4}(V_{pp\sigma} - V_{pp\pi})f_2 & 0 \\ -\tfrac{\sqrt{3}}{2} V_{sp\sigma} f_2 & -\tfrac{\sqrt{3}}{4}(V_{pp\sigma} - V_{pp\pi})f_2 & V_{pp\pi}(e^{ikn_1} + \tfrac{1}{4}f_1) + \tfrac{3}{4}V_{pp\sigma}f_1 & 0 \\ 0 & 0 & 0 & V_{pp\pi} f_0 \end{pmatrix}$$

where:

$f_0 = e^{ikn_1} + e^{ikn_2} + e^{ikn_3}$

$f_1 = e^{ikn_2} + e^{ikn_3}$

$f_2 = e^{ikn_2} - e^{ikn_3}$

Useful Ref: PRB **74** 165310 (2006)

Using same k-point definitions and plotting commands as before, we see a complete bandgap in the sp^2-like bands:

```
Es=-7.3;
Ep=0.0;
Vsss=-4.8;
Vsps=4.75;
Vpps=4.39;
Vppp=-2.56;

H11=diag([Es Ep Ep Ep]);
for kk=1:length(ks)
   k=ks(kk,:);
   f0=exp(i*k*n1')+exp(i*k*n2')+exp(i*k*n3');
   f1=exp(i*k*n2')+exp(i*k*n3');
   f2=exp(i*k*n2')-exp(i*k*n3');

H12=[Vsss*f0 Vsps*(exp(i*k*n1')-0.5*f1) Vsps*sqrt(3)/2*f2 0; ...
    -Vsps*(exp(i*k*n1')-0.5*f1) Vpps*(exp(i*k*n1')+1/4*f1)+3/4*Vppp*f1 -sqrt(3)/4*(Vpps-Vppp)*f2 0 ; ...
    -Vsps*sqrt(3)/2*f2 -sqrt(3)/4*(Vpps-Vppp)*f2 Vppp*(exp(i*k*n1')+1/4*f1)+3/4*Vpps*f1 0   ;...
    0 0 0 Vppp*f0];

   H=[H11 H12; H12' H11];
   eigH(:,kk)=eig(H);
end
```

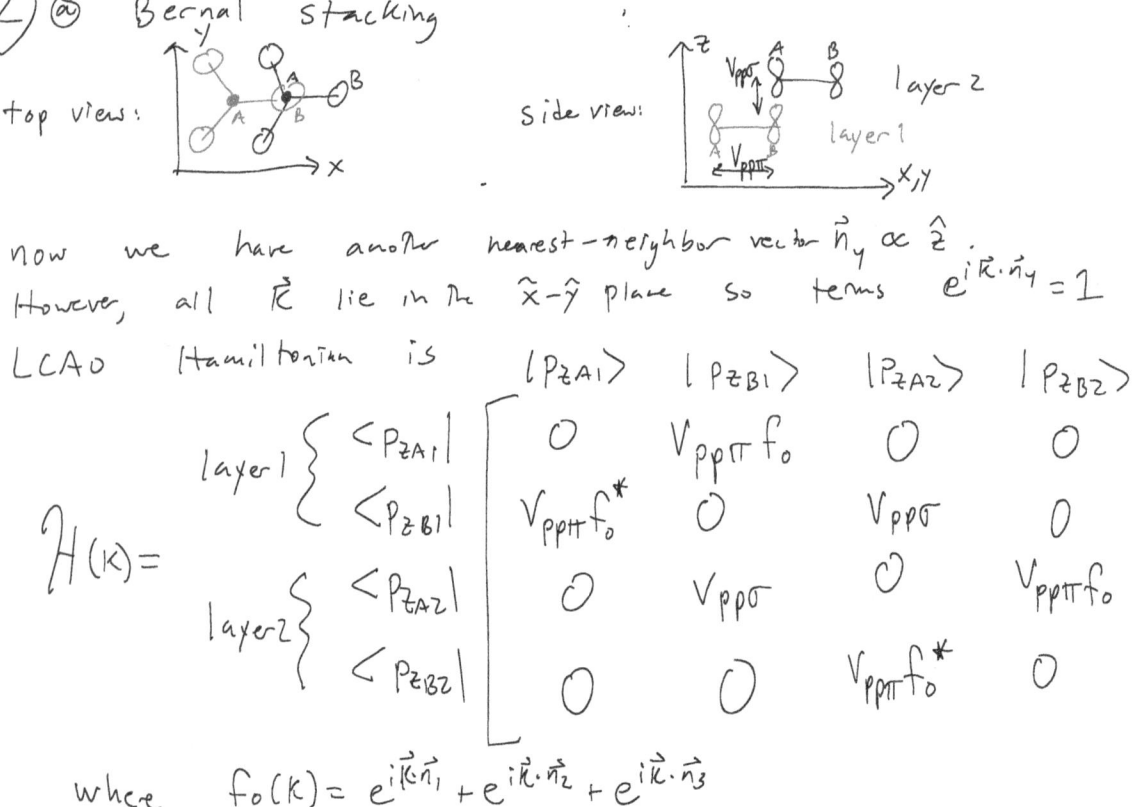

Since the p_z bands are unaffected and E_F is still at Dirac point in complete sp^2 bandgap, we were justified in using our initial 2×2 p_z-only Hamiltonian!

2) @ Bernal stacking

top view:

side view:

Now we have another nearest-neighbor vector $\vec{n}_4 \propto \hat{z}$. However, all $\vec{k}$ lie in the $\hat{x}-\hat{y}$ plane so terms $e^{i\vec{k}\cdot\vec{n}_4} = 1$

LCAO Hamiltonian is

$$\mathcal{H}(k) = \begin{array}{r} \text{layer 1} \left\{ \begin{array}{c} \langle p_{zA1}| \\ \langle p_{zB1}| \end{array} \right. \\ \text{layer 2} \left\{ \begin{array}{c} \langle p_{zA2}| \\ \langle p_{zB2}| \end{array} \right. \end{array} \begin{array}{cccc} |p_{zA1}\rangle & |p_{zB1}\rangle & |p_{zA2}\rangle & |p_{zB2}\rangle \\ \left[\begin{array}{cccc} 0 & V_{pp\pi} f_0 & 0 & 0 \\ V_{pp\pi} f_0^* & 0 & V_{pp\sigma} & 0 \\ 0 & V_{pp\sigma} & 0 & V_{pp\pi} f_0 \\ 0 & 0 & V_{pp\pi} f_0^* & 0 \end{array} \right] \end{array}$$

where $f_0(k) = e^{i\vec{k}\cdot\vec{n}_1} + e^{i\vec{k}\cdot\vec{n}_2} + e^{i\vec{k}\cdot\vec{n}_3}$

2b) $f_0(\vec{K} = (0, \frac{4\pi}{3a}, 0)) = 0$ so

$$\mathcal{H}(\vec{K}) = \begin{bmatrix} 0 & 0 & 0 & 0 \\ 0 & 0 & V_{pp\sigma} & 0 \\ 0 & V_{pp\sigma} & 0 & 0 \\ 0 & 0 & 0 & 0 \end{bmatrix}$$

This has eigenvalues determined by characteristic eqn.

$$\det(\mathcal{H}(\vec{K}) - E\mathbb{I}_4) = 0$$

$$E^2(E^2 - V_{pp\sigma}^2) = 0 \implies E = 0, 0, V_{pp\sigma}, -V_{pp\sigma}$$

2c) $f(\vec{K} + \vec{\Delta k}) = e^{i(\Delta k_x \hat{x} + (\frac{4\pi}{3a} + \Delta k_y)\hat{y})} \cdot \begin{cases} \vec{n}_1 = \frac{a}{\sqrt{3}}\hat{x} \\ \vec{n}_2 = -\frac{a}{2\sqrt{3}}\hat{x} + \frac{a}{2}\hat{y} \\ \vec{n}_3 = -\frac{a}{2\sqrt{3}}\hat{x} - \frac{a}{2}\hat{y} \end{cases}$ (+)

$= e^{i\frac{\Delta k_x a}{\sqrt{3}}} + e^{i(\frac{2\pi}{3} + \frac{\Delta k_y a}{2} - \frac{\Delta k_x a}{2\sqrt{3}})} + e^{i(-\frac{2\pi}{3} - \frac{\Delta k_y a}{2} - \frac{\Delta k_x a}{2\sqrt{3}})}$

$= e^{i\frac{\Delta k_x a}{\sqrt{3}}} + 2e^{-i\frac{\Delta k_x a}{2\sqrt{3}}} \cos(\frac{2\pi}{3} + \frac{\Delta k_y a}{2})$

$= e^{i\frac{\Delta k_x a}{\sqrt{3}}} + 2e^{-i\frac{\Delta k_x a}{2\sqrt{3}}}(\cos\frac{2\pi}{3}\cos\frac{\Delta k_y a}{2} - \sin\frac{2\pi}{3}\sin\frac{\Delta k_y a}{2})$

$\approx 1 + i\frac{\Delta k_x a}{\sqrt{3}} + 2(1 - i\frac{\Delta k_x a}{2\sqrt{3}})(-\frac{1}{2}\cdot 1 - \frac{\sqrt{3}}{2}\frac{\Delta k_y a}{2})$

$\approx i\frac{3\Delta k_x a}{2\sqrt{3}} - \frac{\sqrt{3}\Delta k_y a}{2} = \frac{\sqrt{3}a}{2}(i\Delta k_x - \Delta k_y)$

$$\mathcal{H}(\vec{K} + \vec{\Delta k}) = \begin{bmatrix} 0 & X & 0 & 0 \\ X & 0 & V_{pp\sigma} & 0 \\ 0 & V_{pp\sigma} & 0 & X \\ 0 & 0 & X & 0 \end{bmatrix}$$ where $X \equiv -\frac{\sqrt{3}a}{2}V_{pp\pi}(i\Delta k_x - \Delta k_y)$

2d) Eigenvalues of Hamiltonian above are given by characteristic eq.

$$-E(-E(E^2-x^2) - V_{pp\sigma}(-V_{pp\sigma}E) - x(x(E^2-x^2)) = 0$$

$$E^4 - E^2 x^2 - V_{pp\sigma}^2 E^2 - x^2 E^2 + x^4 = 0$$

$$(E^2)^2 - (2x^2 + V_{pp\sigma}^2) E^2 + x^4 = 0 \quad \Longleftarrow \text{Can use quadratic formula to solve for } E^2!$$

$$E^2 = \frac{2x^2 + V_{pp\sigma}^2 \pm \sqrt{(2x^2 + V_{pp\sigma}^2)^2 - 4x^4}}{2} = \frac{2x^2 + V_{pp\sigma}^2 \pm \sqrt{4x^2 V_{pp\sigma}^2 + V_{pp\sigma}^4}}{2}$$

$$= \frac{2x^2 + V_{pp\sigma}^2 \pm V_{pp\sigma}^2 \sqrt{1 + \frac{4x^2}{V_{pp\sigma}^2}}}{2} \approx \frac{2x^2 + V_{pp\sigma}^2 \pm V_{pp\sigma}^2 \left(1 + \frac{2x^2}{V_{pp\sigma}^2} - \frac{16x^4}{8 V_{pp\sigma}^4}\right)}{2}$$

for $(-)$, $E^2 \simeq \frac{x^4}{V_{pp\sigma}^2}$, for $(+)$, $E^2 \simeq V_{pp\sigma}^2 + 2x^2$

Since $x \propto \Delta k$, we have the following (massive) spectrum:

$E(k) \approx \sqrt{V_{pp\sigma}^2 + 2x^2}$

$= V_{pp\sigma}\left(1 + \frac{2x^2}{V_{pp\sigma}^2}\right)^{1/2}$

$\approx V_{pp\sigma}\left(1 + \frac{x^2}{V_{pp\sigma}^2}\right)$

$\approx V_{pp\sigma} + \frac{x^2}{V_{pp\sigma}}$

$E(k) \approx \frac{x^2}{V_{pp\sigma}}$

$x^2 = \Delta k_x^2 + \Delta k_y^2$

2e) $m^* = \frac{\hbar^2}{d^2 E/dk^2} = \frac{\hbar^2}{\frac{d^2}{dk^2}\left(\frac{3/4 a^2 \Delta k^2 V_{pp\pi}^2}{V_{pp\sigma}}\right)} = \frac{2 \hbar^2 V_{pp\sigma}}{3 a^2 V_{pp\pi}^2} \sim \frac{2 \cdot 0.2 eV \cdot (6.6 \times 10^{-16} eVs)^2}{3 \cdot (2.5 \times 10^{-8} cm)^2 (3 eV)^2}$

$\simeq 10^{-17} eV \frac{s^2}{cm^2}$

In terms of electron mass $m_0 \sim 5 \times 10^5 eV/c^2 \sim 5 \times 10^{-16} eV \frac{s^2}{cm^2}$, this is $m^* \sim 0.02 m_0$

2f) Add $H' = e\phi(z)$ which is diagonal in the tight-binding spatial basis:

$$H' = \frac{e\phi}{2}\begin{bmatrix} -1 & 0 & 0 & 0 \\ 0 & -1 & 0 & 0 \\ 0 & 0 & 1 & 0 \\ 0 & 0 & 0 & 1 \end{bmatrix}$$

2g) energy eigenvalues at degeneracy point K are now roots of characteristic eqn

$$\det\begin{pmatrix} -E - \frac{e\phi}{2} & 0 & 0 & 0 \\ 0 & -E - \frac{e\phi}{2} & V_{pp\sigma} & 0 \\ 0 & V_{pp\sigma} & -E + \frac{e\phi}{2} & 0 \\ 0 & 0 & 0 & -E + \frac{e\phi}{2} \end{pmatrix} = \left(-E - \frac{e\phi}{2}\right)\left[\left(-E - \frac{e\phi}{2}\right)\left(-E + \frac{e\phi}{2}\right)^2 - V_{pp\sigma}\left(V_{pp\sigma}\left(-E + \frac{e\phi}{2}\right)\right)\right] = 0$$

$$= \left(-E - \frac{e\phi}{2}\right)^2 \left(-E + \frac{e\phi}{2}\right)^2 - V_{pp\sigma}^2 \left(E^2 - \left(\frac{e\phi}{2}\right)^2\right) = 0$$

Now, roots $E=0$ are split into $E = \pm \frac{e\phi}{2}$

3) Silicon bandstructure

```
function [Energies, ks]=silicon

Ep=0;
Es=-7.2;
Vss=-8.13;
Vsp=5.88;
Vxx=3.17;
Vxy=7.51;

d1 = [1 1 1]*(1/4);
d2 = [1 -1 -1]*(1/4);
d3 = [-1 1 -1]*(1/4);
d4 = [-1 -1 1]*(1/4);

H11=diag([Es Ep Ep Ep]);
H22=H11;

N=300; %number of k points along path
[ks, kpath, kpathstring]=getks(N);

for ii=1:length(ks)
  k=4*pi*ks(ii,:);
  g1 = (1/4)*(exp(i*(d1*k')) + exp(i*(d2*k')) + exp(i*(d3*k')) + exp(i*(d4*k')));
  g2 = (1/4)*(exp(i*(d1*k')) + exp(i*(d2*k')) - exp(i*(d3*k')) - exp(i*(d4*k')));
  g3 = (1/4)*(exp(i*(d1*k')) - exp(i*(d2*k')) + exp(i*(d3*k')) - exp(i*(d4*k')));
  g4 = (1/4)*(exp(i*(d1*k')) - exp(i*(d2*k')) - exp(i*(d3*k')) + exp(i*(d4*k')));

  H12=[Vss*g1  Vsp*g2  Vsp*g3  Vsp*g4;
      -Vsp*g2  Vxx*g1  Vxy*g4  Vxy*g3;
      -Vsp*g3  Vxy*g4  Vxx*g1  Vxy*g2;
      -Vsp*g4  Vxy*g3  Vxy*g2  Vxx*g1];

  H=[H11 H12; H12' H22];

  Energies(:,ii)=eig(H);

end

plotbands(Energies, ks, kpath, kpathstring, N)

end %%%%%%%%%%%%%%%%%%%%%%%%%%
```

```
function [ks, kpath, kpathstring]=getks(N)

%High symmetry points
L=0.25*[1 1 1];
Gamma=[0 0 0];
X=0.5*[1 0 0];
U=0.5*[1 1/4 1/4];
K=[3/8 0 3/8];
W=0.5*[1 1/2 0];

%choose path through k-space along IBZ edges here
kpath=[L; Gamma; X; U; K; Gamma]; kpathstring=char('L', '\Gamma', 'X', 'U', 'K', '\Gamma');
%kpath= [L; Gamma; X; W; Gamma; U; X]; kpathstring= char('L', '\Gamma', 'X', 'W', '\Gamma', 'U', 'X');
ks=[];
for ii=1:(length(kpath)-1)
  thispath=[];
  for jj=1:3
    thispath=[ thispath [linspace(kpath(ii,jj),kpath(ii+1,jj),round(N*norm(kpath(ii,:)-kpath(ii+1,:))))]];
  end
  ks=[ks; thispath];
end

end %%%%%%%%%%%%%%%%%%%%%%%%
```

```
function plotbands(E, ks, kpath, kpathstring,N)
figure;
plot(E','b'); hold on;
ylabel('Energy  [h^2/(2ma^2)]')
set(gca,'XTick',[]) %no x tick marks
Emin=min(min(E));Emax=max(max(E));
axis([1 length(ks) Emin Emax]);

%superimpose vertical lines and labels at symmetry points
pathlength=0;
text(0,Emin-0.5,kpathstring(1,:));
for ii=2:length(kpathstring)
  pathlength(ii)=round(N*norm(kpath(ii,:)-kpath(ii-1,:)));
  text(sum(pathlength), Emin-0.5, kpathstring(ii,:));
end
Vlinesx=repmat(cumsum(pathlength),2,1);
Vlinesy=repmat([Emin; Emax], 1, length(kpathstring));

plot(Vlinesx,Vlinesy,'k'); hold off;
```

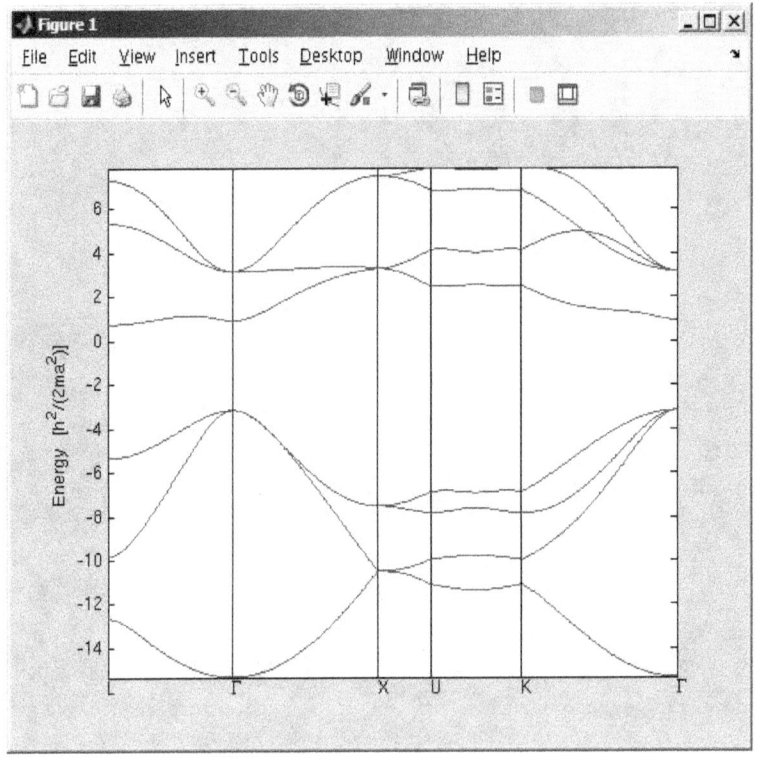

4) Spin-orbit in cubic lattice

```
Lx=1/sqrt(2)*[0 1 0; 1 0 1; 0 1 0];
Ly=1/sqrt(2)*[0 -i 0; i 0 -i; 0 i 0];
Lz=[1 0 0 ; 0 0 0 ; 0 0 -1];

Sx=1/2*[0 1;1 0];
Sy=1/2*[0 -i;i 0];
Sz=1/2*[1 0;0 -1];

LdotS=kron(Lx,Sx)+kron(Ly,Sy)+kron(Lz,Sz);
VSO= 0.5;
HSO(3:8,3:8)=VSO*LdotS; %spin-orbit term in hamiltonian

%construct transformation matrix into definite Lz states from s,px,py,pz basis
sup=[1 0 0 0 0 0 0 0]';
sdn=[0 1 0 0 0 0 0 0]';
mplus1up=[0; 0; -1; 0; -i; 0; 0; 0]/sqrt(2);
mplus1dn=[0; 0; 0; -1; 0; -i; 0; 0]/sqrt(2);
m0up=[0; 0; 0; 0; 0; 0; 1; 0];
m0dn=[0; 0; 0; 0; 0; 0; 0; 1];
mminus1up=[0; 0; 1; 0; -i; 0; 0; 0]/sqrt(2);
mminus1dn=[0; 0; 0; 1; 0; -i; 0; 0]/sqrt(2);
T=[sup sdn mplus1up mplus1dn m0up m0dn mminus1up mminus1dn];

%define path along IBZ perimeter
Gamma=[0 0 0];
R=[pi pi pi];
X=[pi 0 0];
M=[pi pi 0];
kpath=[Gamma; R; X; Gamma; M]; kpathstring=char('\Gamma', 'R', 'X', '\Gamma','M');

N=50; %number of k points from Gamma to K
ks=[];
for ii=1:(length(kpath)-1)
  thispath=[];
  for jj=1:3
    thispath=[ thispath [linspace(kpath(ii,jj),kpath(ii+1,jj),round(N*norm(kpath(ii,:)-kpath(ii+1,:))))']];
  end
  ks=[ks; thispath];
  text(kpath(ii,1)/pi,kpath(ii,2)/pi,kpathstring(ii,:));
end

ks=[ks; Gamma]; %%%%%%%%%%%%%%% added for zone center wavefunction

%nearest-neighbor vectors in cubic lattice
n1=[1 0 0]';
n2=[0 1 0]';
n3=[0 0 1]';

Ep=0;
Es=4;
Vss=0.1;
Vsp=0.1;
Vpps=0.5;
Vppp=0.3;
for kk=1:length(ks)
  k=ks(kk,:);
  H=[Es+Vss*2*(cos(k*n1)+cos(k*n2)+cos(k*n3)) -Vsp*2*i*sin(k*n1) -Vsp*2*i*sin(k*n2) -Vsp*2*i*sin(k*n3); ...%s
    Vsp*2*i*sin(k*n1) Ep+Vpps*2*cos(k*n1)+Vppp*2*(cos(k*n2)+cos(k*n3)) 0 0; ...%px
    Vsp*2*i*sin(k*n2) 0 Ep+Vpps*2*cos(k*n2)+Vppp*2*(cos(k*n1)+cos(k*n3)) 0; ...%py
    Vsp*2*i*sin(k*n3) 0 0 Ep+Vppp*2*(cos(k*n1)+cos(k*n2))+Vpps*2*cos(k*n3)]; ...%pz

  H=kron(H,eye(2)); %add spin
  H=T'*H*T; %transform into basis that diagonalizes Lz (same as HSO)
  Evals(:,kk)=eig(H+HSO);
end

figure(1); %plot bandstructure
Emax=5; TEXTY=0.25;
plot(1:length(ks),Evals); hold on
set(gca,'XTick',[]) %no x tick marks
ylabel('Energy  [eV]');axis([1 length(ks) -Emax Emax]);
pathlength=0;
text(0,-(Emax+TEXTY),kpathstring(1,:));
for ii=2:length(kpath)
  pathlength(ii)=round(N*norm(kpath(ii,:)-kpath(ii-1,:)));
  text(sum(pathlength), -(Emax+TEXTY), kpathstring(ii,:));
end
Vlinesx=repmat(cumsum(pathlength),2,1);
Vlinesy=repmat([-Emax; Emax], 1, length(kpath));
plot(Vlinesx,Vlinesy,'k'); hold off;
```

```
%look at wavefunctions for k=Gamma=[0 0 0]
[v,d]=eig(H+HSO);
Ssqrd=kron(eye(3), Sx)^2+kron(eye(3), Sy)^2+kron(eye(3), Sz)^2;
Lsqrd=kron(Lx, eye(2))^2+kron(Ly, eye(2))^2+kron(Lz, eye(2))^2;
Jsqrd=[Sx^2+Sy^2+Sz^2 zeros(2,6); zeros(6,2) Ssqrd+Lsqrd+2*LdotS];
expctJsqrd=diag(v'*Jsqrd*v);

Jz=kron(Lz, eye(2))+kron(eye(3), Sz);
expctJz=diag(v'*[Sz zeros(2,6); zeros(6,2) Jz]*v);

disp('    E       J^2      |J|      Jz')
disp([diag(d) abs(expctJsqrd) (sqrt(1+4*abs(expctJsqrd))-1)/2 real(expctJz)])
```

Output:

```
>> spinorbit
    E       J^2      |J|      Jz
  1.7000   0.7500   0.5000   0.5000
  1.7000   0.7500   0.5000  -0.5000
  2.4500   3.7500   1.5000   1.5000
  2.4500   3.7500   1.5000  -1.5000
  2.4500   3.7500   1.5000  -0.5000
  2.4500   3.7500   1.5000   0.5000
  4.6000   0.7500   0.5000  -0.5000
  4.6000   0.7500   0.5000   0.5000
```

} spin 1/2 split-off hole
} spin 3/2 { ±3/2 heavy hole } valence band
         { ±1/2 light hole }
} spin 1/2 conduction band

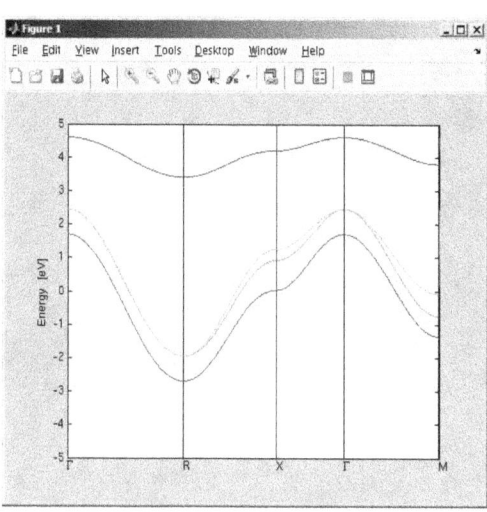

www.ingramcontent.com/pod-product-compliance
Lightning Source LLC
Chambersburg PA
CBHW080918170526

45158CB00008B/2160